Ground Water Quality Protection

STATE AND LOCAL STRATEGIES

Committee on Ground Water Quality Protection

Water Science and Technology Board

Commission on Physical Sciences, Mathematics, and Resources

National Research Council

National Academy Press
Washington, D.C. 1986

National Academy Press, 2101 Constitution Avenue, NW, Washington, DC 20418

NOTICE: The project that is the subject of this report was approved by the Governing Board of the National Research Council, whose members are drawn from the councils of the National Academy of Sciences, the National Academy of Engineering, and the Institute of Medicine. The members of the committee responsible for the report were chosen for their special competences and with regard for appropriate balance.

This report has been reviewed by a group other than the authors, according to procedures approved by a Report Review Committee consisting of members of the National Academy of Sciences, the National Academy of Engineering, and the Institute of Medicine.

The National Research Council was established by the National Academy of Sciences in 1916 to associate the broad community of science and technology with the Academy's purposes of furthering knowledge and of advising the federal government. The Council operates in accordance with general policies determined by the Academy under the authority of its congressional charter of 1863, which establishes the Academy as a private, nonprofit, self-governing membership corporation. The Council has become the principal operating agency of both the National Academy of Sciences and the National Academy of Engineering in the conduct of their services to the government, the public, and the scientific and engineering communities. It is administered jointly by both Academies and the Institute of Medicine. The National Academy of Engineering and the Institute of Medicine were established in 1964 and 1970, respectively, under the charter of the National Academy of Sciences.

Although the information in this document has been funded by the U.S. Environmental Protection Agency under Cooperative Agreement Number CR-811815-01-2 to the National Academy of Sciences, it does not necessarily reflect the views of the agency, and no official endorsement should be inferred.

Library of Congress Cataloging in Publication Data

Ground water quality protection.

Includes index.
1. Water, Underground—United States—Quality.
2. Water, Underground—Government policy—United States.
I. National Research Council (U.S.). Committee on
Ground Water Quality Protection.
TD223.G74 1986 363.7′394 86-12350

ISBN 0-309-03685-2

Printed in the United States of America

COMMITTEE ON GROUND WATER QUALITY PROTECTION

JEROME B. GILBERT, East Bay Municipal Utility District, Oakland, California, *Chairman*
EULA BINGHAM, University of Cincinnati
JOHN J. BOLAND, The Johns Hopkins University
ANTHONY D. CORTESE, Tufts University
THOMAS M. HELLMAN, General Electric, Fairfield, Connecticut
WILEY HORNE, Metropolitan Water District of Southern California, Los Angeles, California
HELEN INGRAM, University of Arizona
THOMAS M. JOHNSON, Levine-Fricke, Inc., Oakland, California
SUE LOFGREN, The Forum, Tempe, Arizona
PAULA MAGNUSON, Geraghty & Miller, Inc., Syosset, New York
PERRY L. McCARTY, Stanford University
DWIGHT F. METZLER, Kansas Department of Health, Topeka (retired)
CHRISTINE SHOEMAKER, Cornell University
DAVID A. STEPHENSON, Dames & Moore, Phoenix, Arizona
JAMES T.B. TRIPP, Environmental Defense Fund, New York

Ex-Officio

DAVID W. MILLER, Geraghty & Miller, Inc., Syosset, New York

Technical Consultant

JOHN B. ROBERTSON, Roy F. Weston Consultants, Washington, D.C.

NRC Project Manager

SHEILA D. DAVID, *Staff Officer*, Water Science and Technology Board

NRC Project Secretary

RENEE HAWKINS, *Senior Secretary*

EPA Project Officers

STEVE CORDLE, Office of Environmental Processes and Effects Research, Washington, D.C.
MARIAN MLAY, Office of Ground Water Protection, Washington, D.C.

WATER SCIENCE AND TECHNOLOGY BOARD

Staff

Preface

In response to a November 1984 request from the Office of Research and Development and the Office of Ground Water Protection of the U.S. Environmental Protection Agency, the National Research Council established the Committee on Ground Water Quality Protection. The committee was asked to identify several state and local ground water protection programs and to review these programs, focusing on prevention of ground water contamination with respect to their scientific bases, performance over time, administrative requirements, and their legal and economic frameworks. The resulting report summarizes the committee's review of case studies and identifies those significant technical and institutional features that show progress and promise in providing protection of ground water quality. It is hoped that these features can be used as practical models for others who are attempting to develop and enhance state and local ground water protection programs.

Chapter 4 contains a number of useful examples that can aid federal, state, and local officials, elected representatives, and citizens in improving ground water protection programs. Its focus is on the future. Today, much of the nation's attention is concerned with correcting ground water degradation resulting from historic practices. However, the nation is facing continued economic development, population growth, and acceleration in the development of new products and technology that make it imperative that governmental programs focus on prevention. The state and regional programs selected for review emphasized planning and regulatory aspects such as information gathering, classification systems, direct and indirect land use controls, and enforcement systems that are preventive rather than corrective. The review of developing or established programs allowed the committee to identify factors that are important in implementing effective programs. Technical and institutional features have also been identified that may have application in other areas of the country. Because, by definition, preventive programs are long range, few explicit results have been demonstrated and thus have not been described in this report. Additionally, the committee did not have the time to review, and was not charged with reviewing, federal policies and programs, that may in some instances limit the effectiveness of state implementation of ground water protection programs.

There are several general policies that can be assumed in the objectives of ground water protection programs. One such policy is strictly economic; that is, the program can be established to meet only cost-effectiveness and related economic objectives. Another approach is oriented to public health objectives, in which the program is structured to achieve health protection goals. A third policy approach is one of total resource management, which addresses all or most interrelated issues and concerns, such as economics, public health, ecological protection, long-term resource conservation, and resource use priorities. The committee assumed a total resource management policy in their approach to examining protection programs. Although some components of the total system receive separate attention in various parts of the report (such as water quality standards or economics in Chapter 4), the committee does not intend to convey an impression that any one of these components should be the principal concern of a protection strategy. Ground water protection programs should encompass all significant aspects of resource management.

The committee, which included experts in water supply management, toxicology, economics, environmental management, ground water quality and protection, political science, hydrogeology, water treatment, civil and environmental engineering, public policy issues, and environmental law, began its deliberations in January 1985. A writing workshop was held in San Diego, California, in July to prepare an initial draft report. Several drafts were produced between July and February 1986, with the final report delivered to the Environmental Protection Agency in April 1986. This study was based on information received from representatives of state and local health and environmental departments or ground water protection offices in the 10 states and 3 local areas chosen for review. The programmatic information and resulting recommendations are based solely on the committee's understanding of the factual situation, which was derived from a brief overview. In-depth study and greater familiarity might change details but not substantive conclusions and recommendations. Those state and local programs reviewed were in Kansas; Arizona; California; Dade County, Florida; the state of Florida; Cape Cod, Massachusetts; the Commonwealth of Massachusetts; Colorado; New York State; Long Island, New York; Connecticut; Wisconsin; and New Jersey. There are many other state and local ground water protection programs and activities, which, unfortunately, could not be reviewed in this study due to time constraints. Many of those programs undoubtedly have worthwhile characteristics that might warrant consideration by anyone contemplating ground water protection strategies.

The committee did not focus on all sources of degradation or contamination that could affect ground water. It studied those activities that have caused significant ground water degradation from a national view. These include solid waste disposal, storage and management of hazardous materials, nonpoint sources of pollution such as septic systems and agricultural chemicals applied to land, land application of waste waters, and production and storage of oil and gas. There are few all-inclusive programs that have been adopted to protect ground water. The committee did not examine how these selected state and local programs address comprehensive hazardous waste management strategies, underground injection of wastes, or materials for oil and gas recovery and mining.

To expand on how well the ground water protection programs were working in these states, the committee conducted personal interviews in several states, in addition to interviewing the state representatives invited to committee meetings. Those interviewed represented not only state and/or environmental health departments but industry, environmental law firms, consulting engineers, and public interest groups among others (see Appendix B). The objective was not to provide an inventory or a nationwide ground water assessment but to identify those ground water protection practices and procedures that have been successful, have experienced difficulty, or might be worth watching in the future. From interviews, discussions, and literature reviews, the committee reached the conclusions and recommendations presented in this report. The report has been jointly researched and written by the committee members and others and reflects a compendium of ideas and thoughts that we hope will aid everyone interested in long-range improvement and protection of ground water quality.

Jerome B. Gilbert, *Chairman*
Committee on Ground Water
Quality Protection

Contents

List of Tables and Figures

TABLES

FIGURES

Ground Water Quality Protection

STATE AND LOCAL STRATEGIES

1

Introduction and Background

The importance of ground water as a source of supply for human consumption, agriculture, and industry and for maintaining surface water ecosystems is now evident. More than 50 percent of the nation's drinking water supply and 80 percent of its rural domestic and livestock water needs are supplied by ground water. According to the U.S. Geological Survey (USGS, 1984), ground water use increased from about 35 billion gallons per day in 1950 to about 87 billion gallons per day in 1980. Withdrawals were expected to approximate 95 billion gallons per day in 1985. This quantity represents about one fifth of the fresh water usage in the nation, the other four fifths being from surface water sources. In contrast, there is approximately 50 times more ground water available at any given time in the United States than there is surface water. The current rate of ground water withdrawal represents approximately 10 percent of the estimated available flow-through of ground water in all aquifers. However, there are many local areas where ground water has been overdeveloped and is being withdrawn at rates in excess of natural replenishment, i.e., mined. During the past decade, growing public concern about ground water quality has resulted from an increasing number of cases of significant contamination or degradation.

In response to these developments, the Congress has enacted several laws that are designed to regulate toxic wastes and substances that are major sources of ground water contamination and to correct some of the most serious contamination problems. These laws include the Safe Drinking Water Act (underground injection wells), the Resource Conservation and Recovery Act (hazardous waste treatment, storage and disposal facilities, and storage tanks for chemicals and petroleum products), and the Comprehensive Environmental Response Compensation and Liability Act (CERCLA or Superfund; uncontrolled hazardous waste and contamination sites). These laws delegate many important responsibilities to the states, which have adopted programs for cleaning up and regulating major sources of ground water contamination.

Many states have broader authority than the federal government to prevent and control ground water contamination. In the last several years, many states and local areas have initiated and expanded ground water protection programs by state mandate. It is hoped that other states can learn from recent experiences in the design and implementation of these ground water protection programs. The committee identified 10 states and 3 local ground water programs in different regions of the country that had potentially effective, and innovative protection features for minimizing and preventing ground water contamination.

No program had all the features of a comprehensive ground water protection program (see "Criteria for Effective Ground Water Programs" below). This is due in part to their formative status; the nature of the state legislative process; the interrelationship of ground water problems with those of other environmental media (air, surface water, and land); and the large variety of human activities that can affect ground water. In most states, different strategies and institutional arrangements for ground and surface waters and land management and permitting activities make it difficult to deal with ground water as a single resource to be protected. This is due in part to the fact that it is not a single resource, isolated from others, but can only be effectively protected by an integrated program that addresses all components of an environment collectively.

The essence of prevention is anticipation, planning, assessment, and preventive action. These preventive efforts anticipate adverse effects from chemical and land use practices and the disposal of waste and provide the necessary protection with emphasis on prevention or control of pollutants at the source. To assist in these efforts, the U.S. Environmental Protection Agency has developed a strategy to protect ground water.

EPA GROUND WATER PROTECTION STRATEGY

The EPA adopted a Ground Water Protection Strategy in 1984 that provides a system for internal coordination as well as a strengthening of state programs. Internal coordination has been improved by issuance of guidelines for EPA decisions affecting ground water protection and cleanup. The guidelines include a three-tiered system for classification of ground water.

Class I is a strict nondegradation category for irreplaceable drinking water supplies and aquifers associated with ecologically vital systems; Class II is current and potential sources of drinking water and waters having other beneficial uses; and Class III is nondrinkable water based on existing poor quality and isolation from drinking water aquifers. EPA accords different levels of protection to each water class. More detailed guidelines on how the classes will be applied are under development.

In the strategy, EPA states its intention to apply its classification system through all of its programs. Where states have already adopted systems for federally regulated sources that are consistent with EPA's approach, the state system may be used. This proposed federal action provides a strong incentive for states to enact their own strategies and classification systems, particularly when a state's overall policies differ from that of EPA. For example, some states have adopted strict nondegradation policies rather than differential protection.

The EPA strategy supports the strengthening of state programs through limited funding and technical support. A $7 million congressional appropriation was made in 1985 for a grant program to be used by states in developing strategies, tools for ground water management, and information collection systems. This funding will support only a small fraction of the cost of implementing comprehensive protection programs and is anticipated to be renewed on an annual basis. EPA has made a commitment to provide research and technical guidance on such issues as the technology for controlling currently unregulated sources such as surface impoundments.

EXTENT OF CONTAMINATION

There are no adequate data available on a national or even regional scale to estimate the extent of ground water contamination and the impacts of this contamination. Some rough estimates, based on oversimplified assumptions, suggest that 1 to 2 percent of our ground water might be contaminated (U.S. Congress, Office of Technology Assessment, 1984), but these estimates may be low because they do not include nonpoint sources of contamination such as pesticide applications. Even though the total percentage of ground water contaminated may be relatively small, it is significant because the areas of contamination generally occur near population centers with a high demand for clean ground water. Over 225 different chemical, radiological, and biological substances have been detected in ground water across the United States (U.S. Congress, Office of Technology Assessment, 1984). There have been numerous cases of ground water contamination across the United States that have caused severe water supply shortages in local areas. Another problem in estimating the extent of contamination is the difficulty in establishing the criteria for defining contamination. There are undoubtedly trace levels of some contaminants in nearly all shallow ground water systems, simply from the percolation and diffusion of airborne contaminants that are ubiquitous in the atmosphere and in rainfall.

According to information provided in the U.S. Congress, Office of Technology Assessment (1984) report and by Pye et al., (1983), the most significant sources of ground water contamination are landfills, surface impoundments, subsurface percolation from septic tanks and cesspools, open dumps, underground storage tanks, and injection wells. The most frequently detected contaminants in ground water generally include chlorinated organic solvents such as trichloroethylene, tetrachloroethylene, trichloroethane, dichloroethane, and dichloroethylene; phthalates and phthalic acid; benzene and ethylbenzene; carbon tetrachloride; chloroform; metals; chloride and nitrate; and radium-226. Many areas have specific problems unique to location and land use activities. This is true for pesticide contamination problems that have occurred in areas such as Florida, the Central Valley of California, and Long Island, New York.

The definition of appropriate management for ground water resources, including the proper role of actions designed to prevent future contamination, requires consideration of the relative advantages and disadvantages of various alternatives.

STATES SELECTED FOR REVIEW

In selecting state and local programs for review, the committee attempted to include those programs that are representative of the range of source types affecting ground water, the hydrogeologic characteristics, contamination problems, institutional arrangement for protection, and the types of protection strategies found across the United States. Where possible, the committee selected those programs that have been in existence long enough to permit some level of analysis of their experience in the programs. Following is a list of the programs selected for review and the principal reasons for their selection:

- Arizona—A state with important quality and quantity issues and a newly established ground water management program.
- California—A large state with a wide range of problems and programs with emphasis on agriculture- and industry-related pollution and an extensive and unique intergovernmental structure for protection programs.
- Colorado—A state with important quality and quantity issues with emphasis on agriculture, residential and commercial development, and mining-related problems and a ground water strategy including a ground water classification system in the beginning stages of implementation.

- Connecticut—A small, densely populated state with a variety of problems and one of the first statewide water classification systems that affects land use and siting of potentially polluting facilities.
- Dade County, Florida—A densely populated and rapidly growing area with a local approach to ground water management including some comprehensive and restrictive laws on source and land use control.
- Florida—A state with special hydrogeologic characteristics generally highly vulnerable to ground water contamination, rapid residential and commercial growth, widespread use of agricultural chemicals, and one that has enacted broad legislation for ground water protection as well as controls on hazardous wastes and deep well injection.
- Kansas—A state with oil production, industrial, and agricultural problems and a long-standing management program.
- Massachusetts—At the state level, an innovative local assistance program including funds for aquifer protection, aquifer mapping, and a classification system.
- Cape Cod, Massachusetts—A fragile hydrologic system, under considerable development pressure. Numerous effective local approaches to ground water protection have been developed and implemented.
- New Jersey—A state with a relatively long-standing monitoring program, and thus well-documented organic chemical contamination problems, a comprehensive industrial and municipal permitting program, and innovative land use controls.
- New York—A large state with a variety of land use and industrial problems, which has completed statewide ground water protection programs.
- Long Island, New York—A densely populated region with unique hydrogeologic characteristics, extensive and diverse ground water contamination sources and problems; a multiagency regional approach to ground water management, and innovative laws on land use controls.
- Wisconsin—A state with a variety of contamination sources, comprehensive statewide controls, and a new multitiered approach to water quality standard setting.

CRITERIA FOR EFFECTIVE GROUND WATER PROGRAMS

The following criteria are considered by the committee to be necessary components of a comprehensive ground water protection program.

1. Goals and Objectives

The goals, objectives, scope, and priorities of a ground water protection program should be clearly defined; they should reflect a comprehensive understanding of the ground water resource problem; and they should be based on adequate legal authority. The program should have objective mechanisms for periodic evaluation of the program's success and a process for achieving needed modifications.

2. Information

A successful program must be founded on an information base that allows proper definition of the resource and the problems, and evaluation of prevention strategies. Many decisions must be made with limited data and scientific uncertainty. However, a prevention program should be based on adequate surveys of (a) water resources and their location, (b) ground water basin characteristics with respect to the potential for contamination, and (c) current and anticipated land and surface water uses that can affect ground water. Information on water resource conditions and new ground water research must be easily available to decision makers.

3. Technical Basis

Effective ground water protection programs require a sound technical basis. Programs should be based on (a) appropriate physical, social, and behavioral assumptions; (b) physical, chemical, engineering, and hydrologic principles; and (c) sound relationships linking mandated actions with desired results. For instance, programs should account for the interconnection of atmosphere, land, and surface water and ground water resources.

4. Source Elimination and Control

In the long term, ground water protection programs should eliminate or reduce the sources of ground water contamination. Simply transferring the problem to another medium such as the air or surface water must be avoided. Sources can be reduced or eliminated by (a) prohibition of certain harmful activities or products; (b) the rational siting of activities and facilities that threaten ground water away from sensitive areas (by land use controls, permits, or regulations); (c) incentives for use of products and technologies less threatening to ground water quality; and (d) incentives for recycling and reuse of waste products.

5. Intergovernmental and Interagency Linkages

Ground water protection programs must link local, state, and federal activities into coherent, coordinated action to be effective. Ground water is affected by a wide variety of human activities and land uses. Moreover, ground water is affected by and, in turn, affects all other environmental media—land, surface water, and air. Consequently, a strong, coherent intergovernmental program is essential in protecting ground water.

6. Effective Implementation and Adequate Funding

Ground water protection programs must have adequate authority and resources and stable institutional structures to be effective. These must include (a) adequate legal authority to take action; (b) adequate long-term funding in light of the seriousness of the ground water problem and efforts necessary to reach objectives; (c) sufficient personnel with adequate training, expertise, and skills and with an ongoing program for professional development; and (d) funding mechanisms and strategies to sustain activity over time.

7. Economic, Social, Political, and Environmental Impacts

A preventive program is based on the assumption that prevention of ground water contamination is the least costly protection strategy in the long run. Therefore, protective actions should be evaluated in terms of their economic, social, political, and environmental impacts. The following factors should be weighed: (a) the feasibility and costs of control, the value of the resource to be protected; (b) the alternative sources, land uses, and environmental and economic impacts that would result from controls; and (c) the potential effects on public health and the environment.

8. Public Support and Responsiveness

Ground water programs must be responsive and credible to interested groups of people and to the general public. Programs should (a) foster public understanding and support, (b) involve the public in program design and evaluation, (c) balance expeditious exercise of authority and consultation with affected parties, and (d) consider the equity of the distribution of benefits, costs, and burdens and the relative ability of various classes and groups of people to bear them.

2
Summary of Conclusions and Recommendations

GENERAL FINDINGS

Ground Water Protection Is Complex and Difficult

Ground water is affected by virtually every activity of society. Numerous activities including industrial production, agriculture, mining, transportation, and commercial and residential activities can contaminate ground water, making development and implementation of effective ground water protection programs difficult. Furthermore, the existence of large and stable governmental programs to protect other environmental media complicates efforts to make ground water protection a priority and to develop an integrated environmental management strategy. Finally, the characteristics and dynamics of ground water systems are more difficult to measure and understand than the more visible and accessible surface and atmospheric resources.

Cleanup of contaminated soils and ground water is difficult and expensive and generally requires many years. Although cleanup activities might be viewed as remedial rather than protective, they can play a significant protective role in preventing contaminants from spreading to clean ground water in the future. The costs, time, and difficulty of major cleanup operations also serve to help emphasize the need for true prevention programs over remediation. The committee recognizes the importance of remedial activities as a ground water protective measure as well as a mitigative one.

However, the emphasis in this report will center on measures other than cleanup programs for ground water protection.

No Single "Comprehensive" Program

While the committee reviewed a number of state programs and noted many promising individual program elements, no single program was found to address all aspects of ground water protection problems comprehensively. A comprehensive program would probably incorporate elements from a number of the state and local programs reviewed in this report as well as other techniques not discussed here. While no single program can be held out as a model for others to follow, collectively they comprise a reasonable array of alternative ground water protection program designs being used in the United States.

State Approaches Differ

It is apparent that ground water issues and conditions vary from state to state and reflect differences in the states' physical, social, and political makeup. As a result, no single strategy for dealing with ground water pollution problems can be recommended for all states or areas. Each state or local strategy will necessarily reflect the local situation.

Importance of Program Focus on Ground Water as a Resource

Programs that base coherent intergovernmental activities on protection of ground water as a resource appear to have a greater potential for long-term success than those that focus on remedial action, as has often been the case in other environmental protection programs.

Overall Effectiveness of Programs Is Unclear

Many programs examined were relatively new and lacked comprehensive criteria and data collection programs to measure their effectiveness. Therefore, the committee could not evaluate their effectiveness in detail. Because ground water protection is a long-term process, success will depend on implementation of effective information-gathering programs. Despite these limitations, it can be seen that many components of the programs examined have clearly been beneficial.

Importance of Increased Priority for Protection as Well as Cleanup

Present and past waste management practices and contamination problems have caused current public perceptions and private priorities to be focused on correction of ground water degradation rather than its prevention. The level of investment in prevention is disproportionately low in comparison to the investment in remedial activities.

Lack of Impact Evaluation

None of the programs studied by the committee was based on explicit evaluation of the health, environmental, economic, social, and political costs and benefits to society associated with the protection of ground water quality. In most cases the information necessary to make such evaluations is not readily available. However, experience with attempts to remedy contamination problems has indicated that lack of prevention is costly.

Lack of Scientific and Technological Information

More scientific and technological information is needed concerning the extent of ground water contamination, its effects on health, the environment, society, and the economy, and strategies and technologies to prevent it. Also, more information is needed on the effectiveness of various protection programs. There is a need for basic research in understanding processes, including those in the unsaturated zone. This is especially true in setting cleanup criteria for specific solutes.

Need for Expanded Federal Action

It is extremely inefficient for each state to fund all scientific data collection and research needed for effective ground water programs. The federal role appears to be inadequate in both magnitude and expertise in the following areas: (1) determination of the health effects of ground water contaminants and establishing drinking water standards; (2) research on source reduction, control methods, and contaminant transport; (3) technology transfer, and (4) exchange of information among states.

Need for Trained Personnel

There appears to be a shortage of appropriately experienced experts in a variety of disciplines in most of the state ground water protection programs reviewed. This also appears to be true in federal programs.

Increased Funding for Protection Programs

All ground water protection programs reviewed by the committee indicated a lack of adequate funding, which constrained the development and implementation of a comprehensive ground water protection program. Although recent efforts through the Clean Water Act and the Safe Drinking Water Act have been helpful in initial planning, increased federal financial support is needed to accelerate and assure the development of self-supporting state and local ground water management programs.

CONCLUSIONS AND RECOMMENDATIONS

Information Base

Hydrogeologic Information

Protection of ground water requires a sound and appropriately designed hydrogeologic information base to determine on a continuing basis what ground water contamination problems may exist. Data are also needed to predict future threats.

The committee encourages state and local programs to obtain the necessary hydrogeological information for each region. The program should be long term to obtain physical and chemical information aimed at developing a quantitative understanding of the occurrence and the quality and dynamics of the resource, together with the types, extent, and sources of potential contaminants. The data should be collected and formatted to assist in the area's ground water management program so that the program's effectiveness over time can be evaluated. The USGS should expand its technical assistance and information-gathering programs to assist states in this effort. State and local organizations should become familiar with and incorporate appropriate data available from federal information systems, such as those of the USGS and the Department of Agriculture, relating to hydrology, soils, and chemical use.

Types of Data and Data Management Systems

Recent advances in electronic data storage and processing technology have enabled the collection and management of large quantities of data. This has often encouraged data gathering without adequate assessment of its usefulness and without conversion of the data into readily usable formats for analyses, policy-making, and management.

The committee recommends that both state and federal information programs be carefully designed to emphasize collection and storage of data that can

be produced in a format that facilitates analysis of problems and long-term trends. The programs should be reviewed and revised regularly to improve the efficiency and selectivity of data gathering. The information management system should be flexible and appropriate to the types and quantities of data anticipated. Data management systems should be easy to access and use but should also be secure from unauthorized manipulation or changes.

Permanent inventory systems for potential contaminants or sources are helpful in preventing ground water and surface water contamination. One such system is the California Pesticide Registry (see Chapter 4), which establishes the quantity, location, and timing of the use of chemicals that could have an effect on water quality.

The committee recommends that states consider establishing ongoing inventories of potential contaminating activities and substances. The compounds and activities inventoried should be selected on the basis of potential risks and quantities of use in each state or regional area. They should include not only traditional sources such as industrial discharges, landfills, and underground storage of chemicals and petroleum products, insecticides, herbicides, fungicides, and fertilizers but also other polluting substances used in significant quantities in land use practices, such as transportation, septic tank cleaning, drilling or mining operations, and underground injection. Such a system can provide valuable information on quantities and locations of substances being used and their potential for contaminating ground water.

Classification

A comprehensive classification system such as that used in Connecticut can be an effective tool for optimizing ground water protection efforts. Maps prepared on the basis of a classification system can be used to guide activities such as the development of standards for water supply, land use management, source controls, and remedial action. By directing the location of potential sources of pollutants away from critical areas, classification can also reduce the cost and controversy associated with case-by-case siting of facilities. In addition, a mechanism for coordination between state and local governments is provided.

Where mapping is not feasible, because of divided authority or data limitations, classification can still provide guidance, especially during permitting and enforcement procedures. However, its usefulness in this case is more reactive than helpful as a planning tool. This is true of the classification systems in Massachusetts and in New Jersey outside the Central Pine Barrens zone, where all fresh ground water is essentially considered to be one class (i.e., drinking water).

The committee recommends that states consider classifying their ground water in conjunction with a mapping program that specifically identifies critical areas and resources for special protection. If data are not sufficient, they should be obtained to provide for classification and mapping in a phased approach. The lack of complete data should not necessarily preclude the development of a classification system. The classification criteria should be adopted through a public process. States with advanced protection programs may opt to give equal protection to all ambient waters of drinking quality.

Comprehensive classification programs depend on adequate hydrogeological information to be effective. Development of the Connecticut classification system was, in large part, due to the existence of historical hydrogeological information produced by the USGS.

The committee recommends that the USGS expand its efforts to produce hydrogeological information to support state and local ground and surface water protection programs.

Ground Water Quality Standards

Water quality standards are set at various levels of government for different purposes. Federal drinking water standards apply to all public drinking water supplies. Additional standards for drinking water at the point of use have been adopted by states.

Depending on their policy, states may also apply numerical standards directly to ground water. These may be designed to protect drinking water quality, other beneficial uses, and critical ecological systems. They may also be used to define nondegradation of high-quality waters. In some cases a safety factor is built in, where the standard is set at a fraction of the enforcement limit. This allows for future growth and the uncertainty associated with ground water protection technology.

One important issue states are faced with is what ambient standards should be set for individual organic contaminants in ground water for various uses, including drinking water, irrigation, and ecological protection. The task of setting these standards is complex. For this reason, most states have looked to the Environmental Protection Agency (EPA) to set standards or guidelines and to provide technical information that states can use in setting ambient standards for pollutants.

Many states have used EPA drinking water standards or recommended limits as a basis for setting ambient ground water quality standards. However, EPA has adopted limits for only a few of the most significant organic chemical contaminants. In the absence of EPA limits, some states such as New York and New Jersey have set ambient and drinking water standards independent of EPA guidance. In late 1984, EPA issued public notice of proposed Recommended Maximum Contaminant Levels (RMCLs) and Maximum Contaminant Levels (MCLs) for many organic and inorganic chemicals in drinking water with appropriate technical supportive information. This information should be immensely helpful to the states in setting standards.

In view of the complexity of setting standards to assist in protection of ground water, the committee recommends the following:

- **EPA should proceed expeditiously with promulgating the RMCLs and MCLs that it has recently proposed, and EPA should propose and promulgate RMCLs and MCLs for all other inorganic and organic chemical compounds commonly found in ground water.**
- **EPA should continue to provide technical information to states about the organic chemicals in ground water for which it has not promulgated RMCLs or MCLs.**
- The application of numerical standards to ground water is a matter of state policy, and there is no single approach that would be appropriate on a national basis. **Ambient ground water standards should be based on the individual state's adopted goals and objectives. These may include protection of beneficial uses other than drinking water, nondegradation, and protection of ecological systems.**
- Wisconsin is one state reviewed by the committee that has developed a two-tiered set of standards designed to limit degradation of ground water and require action by polluters. **In setting standards the states should consider a multi-tiered standard-setting approach that can be used to prevent degradation of high-quality ground water and to protect public health.**
- **In addition, the committee recommends that EPA provide states with a central permanent source of technical information and standard-setting criteria. However, EPA should have the capability to establish overriding standards when states establish inconsistent standards preventing effective long-term prevention of ground water degradation.**

Control of Contamination Sources

Hazardous and Solid Waste Management

Both nonhazardous and hazardous wastes disposed of on land can be major contributors to ground water degradation. While the quantities of waste needing disposal in landfills can be significantly reduced by a variety of source reduction and treatment techniques such as incineration and resource recovery, such processes also generate atmospheric pollutants and ash residues requiring proper treatment and disposal.

The committee recommends that states and communities consider such methods of reducing waste quantities, but only as part of an integrated environmental management program with monitoring requirements, discharge or emission limits, and ambient environmental quality standards for both ground water and air resources that use comparable concepts of risk assessment.

Some of the ground water protection programs examined by the committee prohibit the land disposal of hazardous wastes. Many prohibit the underground injection of hazardous wastes. Many of the states reviewed also lacked other approved methods for hazardous waste management and treatment. These prohibitions and shortcomings usually resulted in exportation of hazardous wastes to other states allowing for their treatment or disposal.

Therefore, the committee recommends that, as an essential element of each state's ground water protection program, a plan should be developed for treating, storing, or disposing of hazardous waste within its boundaries. A program for waste minimization should be a key element in the plan. Such a program should also include a siting process for transportation, storage, and disposal facilities, including regional and on-site industrial incinerators. Exportation of hazardous waste, a temporary expedient that generally increases risks associated with transportation and decreases the assured overall level of environmental protection, should be considered or continued only in special circumstances. The federal government should have a role in mediating this decision in case arbitration is needed.

One of the unique approaches to cleanup and control of hazardous wastes is New Jersey's Environmental Cleanup Responsibility Act of 1983 (ECRA), which places responsibility for cleanup on industry before sale of property to a new owner. This type of legislation could provide an effective prevention incentive as well as remedial pollution control program.

States should consider adoption of programs comparable to the New Jersey Environmental Cleanup Responsibility Act (ECRA) program with broad application to provide incentives for good housekeeping by individual industrial firms, and other significant potential polluting activities.

Underground Storage Tanks

Many local programs such as those in Long Island, New York; Dade County, Florida; and Cape Cod, Massachusetts, and some states, have regulatory management programs for underground storage tanks containing hazardous materials and petroleum products. The committee believes that these programs are effective in reducing ground water contamination from these sources.

The committee recommends that each state should consider developing a comprehensive program for monitoring and inspecting chemical and petroleum product storage tanks with stringent design standards for all new tanks and a requirement for monitoring, testing, and upgrading existing tanks in important recharge areas.

Nonpoint Sources

The purposeful application of agricultural chemicals to land is distinct from most other sources of ground water contamination. Most of these chemicals are pesticides designed to be toxic to insect, fungal, and plant pests and are lawfully released into the environment for beneficial reasons. Besides ground water contamination, pesticides pose a threat to workers and to air and surface water quality.

The task of preventing pesticide contamination is made more difficult by the large number of pesticides in use and the wide range of chemicals and toxicological and environmental-fate characteristics displayed by these materials and the lack of information on their environmental fate and health effects. Thus, the committee has chosen to concentrate its recommendations on nonpoint source contaminants in this area.

- ***Pesticide Use Data Base*** **States should maintain a data base on the spatial and temporal distribution of applied pesticides. Applicators could be required to report when, where, and how much of each pesticide is applied. Also useful are maps and summaries to indicate where such materials were applied.**
- ***Registration Procedure for Certain Chemicals*** **States should initiate a routine procedure for flagging pesticides that have potential for leaching into and contaminating ground water. Such a procedure should be based on the pesticide's chemical characteristics and other factors such as evidence of previous detection in ground water. States should consider canceling the registration of pesticides for which essential data have not been provided.**
- ***Pesticide Tax to Fund Monitoring*** **Monitoring should be used to ensure that currently registered, potentially leachable pesticides do not reach ground water. States should consider funding through fees paid for pesticides or their use. Such a program has two advantages: (1) a cost more reflective of the true cost of the pesticide is then paid by its users, and (2) there is economic incentive for the manufacturers to produce new pesticides that do not have the potential to leach into ground water.**
- ***Cancellation of Pesticide Registration in Local Areas*** **States that are reluctant to cancel the statewide registration for a potentially leachable pesticide should consider canceling registration in local areas where soil conditions or other factors indicate that pesticide leaching may be a serious problem.**

• ***Economic Incentives, Legislation, and Financial Support for Source Reduction*** **States should encourage the use and development of source reduction techniques such as pesticide substitution, changes in irrigation practices, prevention of pesticide and fertilizer application near drinking water wells, and integrated pest management.**

Source Reduction

The committee believes that the best long-term strategy for ground water protection is to reduce and/or eliminate the sources of contamination. There are cases for which source reduction offers an efficient and cost-effective means of minimizing ground water pollution. In general, the state programs examined are weak in source reduction programs such as incineration, recycling, and best management practices; it is evident that additional incentives and information are needed to accelerate and expand source reduction efforts by industry and the public. Therefore, the committee recommends the following:

• **States should consider regulatory and economic incentives for source reduction by industry, government, commercial interests, and the public. States should also consider a variety of financial assistance programs to encourage waste reduction by industry, such as low-interest or no-interest government loans for capital cost of new equipment or environmental audits to determine the best way of reducing waste generation; tax reductions or credits; grants or other aid to encourage smaller firms to pool resources and implement a joint waste reduction strategy; government subsidies to firms actively working on new methods of reducing wastes; financial assistance to waste exchanges to encourage more recycling and reuse of materials that might otherwise be disposed of on the land or into ground water.**

• **State agencies, university-based groups, trade associations, and other institutions should develop educational programs for local industries and the public to disseminate information on waste reduction technology and assist them in implementing waste reduction practices. Specific emphasis should be on medium- and small-sized generators of industrial waste that do not have the expertise or time to keep abreast of technological opportunities.**

• **The committee believes that EPA should fund additional research on source reduction technologies. EPA should also fund programs that include research into public and private practices in the use of substances that are potential ground water contaminants.**

Prohibition of polluting activities is one of the most effective means of source reduction. This includes eliminating ground and underground discharges, banning the use of potentially polluting products, and prohibiting certain activities in important ground water recharge areas. Several areas of the country have used prohibition of polluting activities successfully through source regulation or land use controls. For example, Suffolk County (New York) and the state of Connecticut have banned the use of organic septic system cleaners. Florida and the New Jersey Pinelands Commission prohibit land disposal of hazardous waste. These and other programs that have used prohibition are discussed in Chapter 4.

The committee found that several state and local entities use hazardous and solid waste disposal strategies other than land disposal. For example, programs such as Florida's "Amnesty Days" have successfully collected household hazardous waste and hazardous waste from small-quantity industrial and commercial generators that otherwise might have been improperly disposed of. However, few states or communities have adopted programs for aggressively promoting source reduction of hazardous waste and recycling of solid waste.

The committee therefore recommends that all states and local entities consider similar strategies for reducing improper disposal of household and other small-quantity generator hazardous waste. Municipalities and states should also consider the relative merits of comprehensive solid waste recycling and incineration programs.

Land Use Controls

Almost every human activity has some potential for contaminating the underlying ground water. The degree of risk is determined by hydrogeologic conditions at a given location and characterization and use patterns of the potential pollutants. Restrictions on land use activities in certain designated sensitive areas can be a significant component of a ground water protection program and may be linked to a ground water classification system.

Land use control is a good complement to source control programs and can significantly increase the level of protection and reduce the cost of both programs. The land use control programs reviewed were implemented at the county or municipal level (i.e., Massachusetts, New Jersey, New York, Connecticut, Florida). However, projects described in this report (Chapter 4) have demonstrated that effective planning at one level of government can lead to implementation at various levels, ranging from towns to state jurisdictions. If provided with planning and technical support from the state and EPA, many localities can develop effective ground water protection programs employing land use controls.

The committee recommends that land use controls be considered an essential part of a ground water protection program. Although land use controls are best

carried out at the local level, state governments can encourage land use controls in combination with other measures to protect ground water. The effectiveness of land use controls is limited by preexisting development. Therefore, land use controls should be implemented at early stages for vulnerable undeveloped areas. Although information on land use conditions is important, it is neither necessary nor possible to have sufficient data to answer all concerns before enactment of protective ordinances. These controls can be revised as new data are gathered.

Implementation of Ground Water Protection Programs

Successful ground water protection programs require adequate legal authority and substantial funding for planning and design as well as implementation. Other factors affecting the successful implementation of ground water protection programs include the tractability of the problem, the size of the target group whose behavior is to be changed, the extent of behavioral change required, the degree of integration within and among implementing institutions, the amount of media attention directed toward the problem, and the commitment and leadership skills of implementing officials. The committee would have liked to go into further detail concerning the difficulties that may be involved in the eventual implementation of policies designed to protect ground water, but due to time limitations the members could discuss only a limited number of these issues. Many of the more attractive programs examined—for example, California, Long Island, New York, and Cape Cod, Massachusetts—have benefited from past federal support under Sections 106 and 208 of the Clean Water Act and the Safe Drinking Water Act.

- **The committee recommends that the federal government provide financial support for development and implementation of state- or basin-level programs on the condition that within a specific time period the states are committed to develop self-supporting ground water management programs.**
- **Long-term program success requires adequate and continuing funding. This is necessary to maintain a strong regulatory surveillance and enforcement effort with substantial information collection and analytical support. States should consider a variety of funding mechanisms including user and disposal fees as well as general revenues for program support.**
- **States should play a key role in expanding the number of well-trained hydrogeologists by providing more support of hydrogeologic programs within the universities and colleges.**
- **The federal government should also provide technical support to state and local governments through research on health and environmental effects of**

ground water contamination, fate and transport of pollutants, and technologies and strategies for ground water protection. The federal government should also establish criteria, guidelines, and standards for important ground water contaminants to ensure national consistency and avoid duplication of efforts among states. In addition, the federal government should provide training of state and local officials in ground water management and protection.

Political Mobilization, Public Participation, and Support

Successful ground water protection programs emerge from circumstances where political support is mobilized for the passage and implementation of effective policies. Ground water degradation has been slow in emerging as a major health and environmental issue because the resource is generally invisible to the public; it is linked in complex ways to land and other resource uses. Causes and effects of pollution are hard to identify and poorly understood. In order for the ground water issue to take a high-priority position on the public agenda, the issue needs to be perceived as a matter of broad social concern as well as one requiring technical resolution. To facilitate political mobilization, public participation, and support, the committee recommends the following at every governmental level:

- **Decision-making processes concerning ground water should be characterized by openness, should reflect consideration of public attitudes, and should include active participation of public health and environmental interest groups, industry, and the public.**
- **Attention should be directed to the need to attract and develop high-level political leadership to shepherd ground water protection legislation and ensure commitment to continued funding and implementation of ground water programs.**
- **Communication networks must be established and maintained between ground water program managers and the media. Media coverage of ground water issues is more likely to be fair and balanced when managers have established a reputation for openness and accuracy. The scientific community should also share responsibility for assisting in dissemination of clear, accurate, and understandable information by the media.**
- **The sharing and exchange of information regarding ground water protection problems and programs for their resolution should be an ongoing component of every program. This may be achieved through various activities and mechanisms, including regular community meetings, workshops, and symposia that provide full opportunity for discussion, reaction, and recommendation by the interested community concerning the program and issues.**
- **Ongoing educational activities about ground water in the context of envi-**

ronmental protection should be undertaken in the school system at all grade levels.

- **States should play a key role in expanding the number of well-trained hydrogeologists by providing more support of hydrogeologic programs within state universities and colleges.**
- **A public intervener-type program should be considered when public confidence or interest is not recognized or adequately incorporated in ground water protection programs. A program such as those in Wisconsin and New Jersey can provide the public with an avenue for legal action to address a perceived problem, and at the same time prevent nonmeritorious suits from being filed against an agency. The public intervener should not be subject to political pressure or changes in administrative policy. An independent public advisory group could help to screen the actions to be taken.**

Role of Economic Analysis

Effective ground water protection programs have significant costs associated with them that can, in some cases, exceed the value of the resource or the costs of remedial actions. While analytical techniques are evolving rapidly and data bases are growing, significant application difficulties remain. These difficulties will not be removed until serious attempts are made to perform economic analyses of ground water protection programs and strategies. Meanwhile, social, political, and economic conditions continue to evolve, shifting costs and values so that it is likely to become more and more difficult to strike the right balance between prevention and remedy, or between universal policies and problem-specific measures.

The committee believes that economic analysis is one of the useful ways programs and strategies can be judged. Economic analyses should be conducted of existing and proposed ground water protection measures so that experience can be gained with techniques and data requirements, and decision-makers can become familiar with the results of such analyses. Such analyses have been performed in connection with hazardous waste cleanup activities (assessment of Superfund natural resource damages, for example) and may be useful in evaluating ground water protection programs.

Comparison of Governmental Ground Water Responsibilities

The developing roles of various levels of government are confusing and, as recommended earlier, should be clarified and expanded in a number of areas. Table 2.1 shows one concept of the relative roles and functions of government that may enhance the reader's understanding of the ways in which the needs identified in this report can be satisfied.

TABLE 2.1 Ground Water Case Study—Various Roles of Government Units for Consideration in the Protection of Ground Water

Governmental Level	Information	Planning	Standards and Enforcement	Public Process
Federal				
USGS	National water quality data, surface/ground.	Conducts local studies or portions of studies jointly with states or local agencies.	Provides technical data on resource conditions to EPA and states.	Provides information from data storage and studies.
EPA	Assesses national ground water conditions and provides technical advice to states.	Provides grants conditioned on establishing self-sufficient continuing planning. Conducts basic national health research.	Establishes national standards and provides technical and scientific basis for state standards. Supports states in enforcement of standards. Initiates national source control.	Is national focus of scientific and public dialogue on ground water protection strategy.

State[a]	Maintains central record of local quality and quantity of ground water, chemical usage, and disposition and present and projected land use.	Conducts continuing statewide planning and provides assistance and grants to regional or local agencies preparing ground water management plans.	Sets state or regional standards, including prohibitions when required national standards have not been set, but based on national scientific assessment. Primary enforcement agency for state and federal standards and source control regulations.	Provides information, education, and continuing forum for public dialogue on goals, standards, control strategies, and investments.
Regional[b]	Provides same information as state for a specific ground water basin serving several local agencies.	Prepares and implements specific basin management plans.	Accepts state delegation for enforcement as appropriate, including standard setting source controls, class systems, and land use plans to reflect actual environmental conditions.	Is primary focus of public process, including citizens, committees, working groups, and information gathering and dissemination.
Local	Provides basic information to individual users.	Prepares and implements local ground water plans.	Enforces appropriate standards. Assists state or regional entities in specific local situations.	Is primary contact with individual citizens.

[a]Each state's structure varies; ground water protection responsibilities may also vary; clear central state coordination is needed.
[b]Interstate Basin Agency: coordinates roles in several states where ground and related surface water problems exist.

3

Summaries of the State and Local Ground Water Programs Reviewed

This chapter provides a brief overview of the ground water conditions and protection programs for each of the state and local areas selected by the committee for in-depth review: Arizona, California, Colorado, Connecticut, Florida and Dade County, Kansas, Massachusetts and Cape Cod, New Jersey, New York and Long Island, and Wisconsin. These overviews will provide a perspective on the characteristics and importance of the ground water resources for each of these areas, a description of the principal ground water quality issues within each area, and a summary description of the area's ground water management and protection programs. These summaries give the background and lay the framework for the more detailed descriptions and evaluations of various protection program components provided in Chapter 4.

This information has been taken from many sources, including the U.S. Geological Survey's (1984) *National Water Summary 1984; Protecting the Nation's Groundwater from Contamination*, Volumes I and II, by the U.S. Congress, Office of Technology Assessment (1984); *Overview of State Ground-Water Program Summaries*, Volume 1, by the U.S. Environmental Protection Agency's Office of Ground-Water Protection (1985b); information given to the committee by state health department representatives at committee meetings; and information obtained through individual interviews conducted by committee members in several states.

ARIZONA

Overview of Ground Water Resources

Availability of adequate and potable water supplies in Arizona has had a great effect on the location of cities and farmlands. Because evapotranspiration greatly exceeds rainfall, agriculture depends almost entirely on irrigation. However, the amount of surface water available is not sufficient to meet the increasing demands. Therefore, ground water reservoirs are of prime importance as a source of water. In 1980, about 58 percent of the total water supply in the state came from its ground water reservoirs, creating an annual overdraft of 2.5 million acre-feet.

The principal use of ground water is for irrigation of crops, although municipal and industrial uses are increasing at a rapid rate, reflecting the Sun Belt boom. Also, more industrial enterprises are being developed in the state.

Arizona's principal aquifers consist of unconsolidated alluvium, consolidated sedimentary rocks, and crystalline igneous and metamorphic rocks. Arizona is divided into three water provinces: the Colorado Plateau uplands province in the northern part of the state, the Basin and Range lowlands province in the southern part of the state, and the Central highlands province, which is transitional between the other two provinces. Table 3.1 describes Arizona's aquifers according to the water province in which they occur from youngest to oldest.

Ground Water Quality Issues

Major sources of ground water contamination in the state are pesticides, fertilizers, irrigation return flows, sanitary landfills, industrial solvents, mining activities, injection wells, surface impoundments, and leaking underground storage tanks and pipelines. Other ground water pollution problems are disposal of waste waters, disposal of waste treatment by-product, salt water intrusion, and spills. Recently, the Arizona Department of Health Services established action levels for 27 volatile organic compounds (VOC) and 3 pesticides. These 30 chemicals, now with action levels, are those particularly relevant to ground water quality in Arizona. Twelve additional VOCs are currently under study for action level designation. The major chemical of concern currently is trichloroethylene (TCE). The action level for TCE is 5 parts per billion (ppb).

TABLE 3.1 Aquifer and Well Characteristics in Arizona

Aquifer name and description	Well characteristics			Remarks
	Depth, common range (ft)	Yield (gal/min)		
		Common range	May exceed	
Alluvial aquifers: Generally sand, gravel, silt, and clay. Occur in the Basin and Range lowlands and parts of the Central highlands. Confined and unconfined	100 - 2,000	1,000	2,500	Thickness from a few hundred to about 10,000 ft. Deposits grade in texture from large boulders near mountains to fine-grained sediments along axis of valleys. In places, dense clay beds form confining layers for permeable sand and gravel beds beneath. Provides water for most cities and extensive irrigated areas in southern part of State.
Sandstone aquifers: Mostly fine-grained sandstone units; fracturing and faulting increases permeability; in places, siltstone and claystone layers function as confining beds. Occur in parts of the Central highlands and in the Plateau uplands. Confined and unconfined.	50 - 2,000	0 - 50	500	Thickness from about 200 to 500 ft. Aquifers may be as much as 1,000 ft below land surface and are separated by thick, relatively impervious layers. Coconino and Navajo Sandstones provide largest supply of water for all uses in central and northern parts of State.
Low-yielding bedrock aquifers: Crystalline and sedimentary rocks. Permeable only where extensively fractured and faulted. Confined and unconfined.	50 - 1,000	0.5 - 2	200	These rocks are generally not considered to be aquifers but do supply usable quantities of water to individual sources for domestic supply in rural areas.

SOURCE: U.S. Geological Survey, 1984.

Ground Water Management and Protection

Ground water quality management is included as a priority environmental issue in a 1985 State-EPA Agreement because EPA and Arizona agree that there is a need to coordinate EPA's new evolving National Ground Water Strategy with the state's existing ground water management program.

In May 1986, the Arizona Legislature passed a landmark Environmental Quality Act that provides for strong pollution control and remedial action measures. In addition, several other bills impacting water quality management—e.g., dry wells, storage tanks, and recharge—were passed; these bills strengthen control of adverse practices concerning ground water.

The Arizona Department of Water Resources is responsible for water quantity management. The Arizona Department of Health Services is the designated water quality management agency by both state and federal law. The Arizona Ground Water Management Act, enacted on June 12, 1980, is the first comprehensive legislative framework for managing the ground water resources of the state. Its stated goal is reaching "safe yield" for water-scarce areas of the state. This act created the Department of Water Resources and made it responsible for administering the law's complex provisions. It established four Active Management Areas—areas in which intensive ground water management is needed because of the large and continuous ground water overdraft. All classes of uses must comply with their designated conservation requirement according to the current Water Management Plan. All wells must be registered and metered, and a withdrawal fee is charged for the amount of water withdrawn. The fee is used partially for administration of the department and during future management periods will help to fund water augmentation projects. Another important method used to protect ground water in Arizona is the requirement for ground water discharge permits, which apply to all activities not otherwise regulated from a ground water standpoint such as large septic tanks, mine tailings ponds, and National Pollution Discharge Elimination System (NPDES) discharges. All future water plans must incorporate water quality in the management scheme.

CALIFORNIA

Overview of Ground Water Resources

More than 10 million people in California, or 46 percent of the total population, are served by ground water supplies. Even more significant is the fact that of the 12.5 billion gallons per day of ground water produced, 39 percent is withdrawn for irrigation. The geography and climate of the state are the dominant factors controlling California's water development. The principal centers of population and agriculture are mostly in water-deficient areas. Many of the valleys and plains of these areas are underlain by productive aquifers. The eventual realization that these underground aquifers were not unlimited was an important factor in the decisions that led to the large-scale importation from water-abundant areas of the north to water-deficient areas of the south.

Rainfall is extremely variable in California, with mean annual precipitation ranging from more than 40 inches in the mountainous areas of central and northern California to less than 5 inches in the desert areas. Natural recharge of ground water, from precipitation and stream infiltration, averages about 5.2 billion gallons per day statewide. Ground water is also recharged by an estimated 6.6 billion gallons per day of applied irrigation water, which percolates through the root zone to the water table.

California is one of the most physiographically and geologically diverse states in the United States. The mountains are formed of consolidated sedimentary, metamorphic, and igneous rocks. Geologic structures are complex, with abundant folds and faults, many of which are active. Earthquakes are common, particularly in the Coast Ranges. The valleys of California are filled with alluvium and other sedimentary materials that comprise most of the principal aquifers. About 40 percent of the land in California is underlain by aquifers. These aquifers are composed of alluvium and older sediments, mostly of continental origin, and volcanic rock. The sedimentary aquifers underlie the major valleys, coastal plains, and desert basins. Alluvial and other sedimentary aquifers in the state are divided into four geographic areas: coastal basins, Central Valley, Southern California, and desert areas. The statewide ground water basin identification report published in 1975 by the Department of Water Resources identified 248 ground water basins. Most of these basins have a rather complex hydrogeological structure with a mixture of confined and unconfined aquifers. A summary of aquifer well characteristics is given in Table 3.2.

Ground Water Quality Issues

Major sources of ground water pollution in California are from agricultural activity such as pesticides and fertilizers, hazardous waste sites, and leaking underground storage tanks and pipelines. Other sources of ground water pollution are municipal landfills, surface impoundments, salt water intrusion, open dumps, and disposal of waste waters, spills, accidents, industrial sites, and military installations.

Major ground water issues include ground water overdraft, salt water intrusion, land subsidence, and artificial recharge and conjunctive use of ground water resources. In addition, salt buildup due to agricultural, municipal, and industrial uses and improperly constructed or abandoned wells may be a significant problem.

Ground Water Management and Protection

California does not have statewide comprehensive ground water management laws. Management is practiced largely by local agencies, using powers provided under state enabling legislation. The California Department of Water Resources is the principal state-level water supply and planning agency. Its role in ground water is one of providing advice and technical support to local agencies, collecting data, and conducting investigations.

The responsibility to protect California's ground water against toxic or hazardous waste pollution is divided three ways among the State Water Resources Control Board (SWRCB), the California Department of Health Services (DOHS), and the California Department of Food and Agriculture (DFA). The SWRCB oversees the quality of surface and ground water under the Porter-Cologne Water Quality Control Act. The DOHS controls a number of chemical-handling activities that can have an impact on ground water quality under the California Hazardous Waste Control Act of 1972, the Federal Resource Conservation and Recovery Act of 1976 (RCRA), and state and federal Superfund cleanup activities. The DFA regulates and administers controls on pesticide applications, pursuant to the California Food and Agricultural Code. Local regulation is implemented by the county agricultural commissioners under authority of the DFA director and county ordinances. Pesticide use is also restricted by the Federal Insecticide, Fungicide, and Rodenticide Act (FIFRA), which is administered by the EPA.

SWRCB

The SWRCB protects ground and surface water quality through a series of Water Quality Control Plans (Basin Plans) for the various geographic areas, which are administered by nine regional boards. Each Basin Plan assigns beneficial uses to ground and surface waters and places limits on ambient levels of total dissolved solids, nitrate, chloride, on gross organics by testing their chemical oxygen demand (COD), and others—as needed to protect those beneficial uses. The regional boards place discharge limits on point and nonpoint sources of waste water that could adversely affect surface or ground water. The California Water Code allows the SWRCB to file

TABLE 3.2 Aquifer and Well Characteristics in California

Aquifer name and description	Water withdrawals in 1980 (Mgal/d)	Well characteristics: Depth (ft) Common range	Well characteristics: Depth (ft) May exceed	Well characteristics: Yield (gal/min) Common range	Well characteristics: Yield (gal/min) May exceed	Remarks
Alluvium and older sedimentary aquifers:						
Coastal basins: Sand, gravel, silt, and clay; continental and marine origin. Unconfined and confined.	1,630	50 - 500	1,000	500 - 1,000	3,000	Aquifers consist of alluvium and older sediments that fill valleys which are tributary to the Pacific Ocean. Multiple aquifer systems are common. Most intensively developed areas are in Santa Clara, Salinas, and Santa Maria Valleys and Santa Rosa area.
Southern California: Sand, gravel, silt, and clay; continental and marine origin. Unconfined and confined.	1,720	50 - 1,000	1,500	500 - 1,500	4,000	Productive aquifers in coastal plains and inland valleys of Ventura, Los Angeles, Orange, and San Bernardino Counties. Seawater intrusion, once a problem in coastal areas, now under control.

Central Valley: Sand, gravel, silt, and clay; continental and marine origin. Unconfined and confined.	10,000	50 - 500	1,000	50 - 1,500	3,000	Largest aquifer system and greatest concentration of ground-water pumpage in California. Corcoran Clay Member, an extensive confining layer, exists in much of San Joaquin Valley.
Basin-fill, desert areas: Sand, gravel, silt, and clay, mostly of continental origin. Unconfined and confined.	700	20 - 400	1,000	200 - 1,500	4,000	Aquifers in some basins deep, and some wells have large yields. Recharge limited by little rainfall. Some aquifers recharged by runoff from streams that originate in high mountains.
Volcanic rocks: Andesite, rhyolite, and basalt. Mostly unconfined; confined locally.	unknown	75 - 200	300	100 - 1,000	4,000	Water occurs in rubble zones, pipes, and fractures. Well yields extremely variable, with a few exceptionally productive wells and many dry holes. Potential yield far exceeds present use.

SOURCE: U.S. Geological Survey, 1984.

an action in the Superior Court to prevent degradation to ground water quality. Recently, the SWRCB, acting in coordination with the regional boards and the counties, has been charged with the registration and regulation of underground containers of hazardous substances pursuant to AB 1362 (Sher) and AB 2013 (Cortese). Additional recent legislation, AB 3566 (Katz), requires strict controls on surface impoundments containing hazardous material.

DOHS

The DOHS administers the state's RCRA and Superfund programs, under federal law and under the preceding State Hazardous Waste Control Act of 1972. Under these programs, DOHS regulates activities that could affect ground water—such as toxic waste generation, treatment, storage, and disposal. In addition, pursuant to Title 22 of the California Administrative Code, the DOHS has the responsibility for quality control of drinking water supplies. Most recently, under State Assembly Bill 1803 (Connelly), DOHS has carried out a survey of well contamination on a statewide basis, for water systems containing more than 200 connections. However, under recent legislation, this program has been extended to systems with fewer than 200 connections. When positives are found, results are referred to the regional boards, to locate pollutant sources. Within each land use area (determined by aerial mapping previously carried out by the Corps of Engineers), a test list of pollutants was identified. Typical wells in Southern California, for example, were tested for 20 to 30 pollutants. Survey results to date are based on 2558 wells in 753 water systems. The data indicate that 315 wells in 126 systems reported positive results, of which 115 wells exceeded either a state "action level" or an EPA maximum contaminant level.

Significant Problems

There are three primary categories of toxic wastes being controlled in California: (1) pesticides, more than 90 percent of which are used on farmlands; (2) industrial solvents, heavy metals, acids, cyanide, PCB, and other wastes, which are largely generated in urban areas; and (3) gasoline, other vehicle fuels, and stored chemicals.

In 1980, more than 120 million pounds of pesticide use (concentrated weight) was reported to DFA. A recent report states that the actual use might have been 3 times that amount owing to reporting gaps. The application of pesticides to land is essential to their use. Hence, the "prevention of ground water contamination by pesticides is essentially a matter of not allowing *significant quantities* of pesticides to enter ground water bodies"

(Ramlit Associates, 1984). The remedies include reducing application rates, selecting less persistent and less mobile pesticides, taking care in container disposal and other "housekeeping" activities, placing grout seals on agricultural wells to prevent short-circuiting of chemicals down well casings, and preventing backsiphonage through combined irrigation/chemical injection rigs. All these practices presume the continuing use of pesticides. The key problems faced by agencies regulating pesticide use are lack of monitoring data for pesticides in ground water, lack of knowledge of pesticide mobility and fate in the ground, and lack of knowledge of short- and long-term effects of particular pesticides sufficient to determine action levels (Ramlit Associates, 1984).

In 1983, about 3.5 million tons of urban hazardous wastes (wet weight) were hauled from generator sites off-site to treatment, storage, or disposal sites. Since most of the total tonnage is treated and disposed on-site at industrial locations rather than hauled off-site, the total wet weight of hazardous wastes generated in California apparently exceeded 7 million tons in 1983 (State of California, Toxic Substances Control Division).

The eventual goal of urban hazardous waste management is to treat and neutralize the waste stream before it is placed on land. At present, however, the state is primarily engaged in enforcing RCRA, through a system of prohibitions and storage and disposal regulations, backed by enforcement. The size and scope of the enforcement program is discussed further in Chapter 4. The overriding management problem with respect to urban hazardous wastes in California is the lack of treatment-based neutralization and disposal facilities. The ultimate off-site disposal points for hazardous wastes at present are seven landfills: four "Class I sites" and three "Class II-1 sites." Under the state's landfill classification scheme, Class I sites are the most secure and are supposed to provide geologic containment by means of an impermeable barrier to prevent leachate from reaching underground waters. Class II-1 sites are currently allowed to receive hazardous wastes but are being phased out as they do not provide for geologic containment. A recent report on the subject (Environmental Defense Fund, 1985b) concludes that none of the disposal sites is meeting the established criteria for containment of wastes. The State Water Resources Control Board has recently revised its land disposal regulations. The classification systems for both sites and wastes have been reviewed and made more stringent.

The underground storage tank problem became prominent in the San Francisco Bay area (Silicon Valley) with a significant leak of solvent reported by the facility owner. This experience triggered two pieces of legislation to register and regulate both existing and new underground containers. As a result of the legislation, more than 165,000 underground storage containers have been registered in California: 81 percent are for motor vehicle fuel; 12 percent for chemicals; 4 percent for sumps; and 1 percent for pits, ponds, and lagoons. Recent estimates are that 85 to 98 percent of the eligible tanks have actually been registered (State Water Resources Control Board staff estimates, 1985).

Estimates vary on how many of the underground tanks are leaking. In the San Francisco Bay area, the regional board's study indicated that 60 percent of the high-priority sites had leaks (SWRCB staff estimate, 1985). Overall, it is currently estimated that 20 percent of the facilities have tanks that leak, but that perhaps one half of those over 10 years of age leak. Thus, the problem tends to worsen with time, so that prevention and correction are critical. Details and costs of the underground container program are discussed further, in Chapter 4.

MASSACHUSETTS AND CAPE COD

Overview of Ground Water Resources

Massachusetts

As in many of the surrounding northeastern states, ground water is an important public supply in Massachusetts. Approximately one third of the 5.7 million people in the state obtain their water supply from wells. About 24 percent of municipal public supplies come from ground water, and 100 percent of rural domestic supplies are from ground water. About 30 percent of ground water withdrawals are used for industrial supplies, and about 2 percent of ground water withdrawals are used for irrigation. The remaining 70 percent of ground water withdrawal is for public and private water supplies.

Most of Massachusetts is underlain by crystalline, metamorphic, and igneous rocks that are covered by a discontinuous mantle of glacial till and stratified drift sediments. The principal aquifers in Massachusetts can be grouped according to general rock type and stratified glacial drift, sedimentary bedrock, carbonate rocks, and crystalline and bedrock.

Stratified glacial drift aquifers provide water for virtually all public supplies that use ground water. This discontinuous aquifer series consists of layered sand and gravel with some silt and was deposited over bedrock by glacial melt waters at the end of the Pleistocene glaciation. In the southeastern corner of Massachusetts the stratified drift aquifer forms a continuous layer over bedrock, which covers the entire areas of Cape Cod, Plymouth County, and the islands of Martha's Vineyard and Nantucket. These extensive permeable aquifers with shallow water tables are highly susceptible to contamination. The natural water quality in these stratified drift aquifers is generally good unless it has been degraded by human activities (see Table 3.3).

Cape Cod

Cape Cod receives special attention in this report because it not only has some interesting hydrogeological characteristics but also has some innovative and effective local approaches to ground water protection. The Cape is composed predominantly of highly permeable stratified drift sediments with a shallow water table and ground water of naturally high quality. Because ground water is the only source of drinking water for residents, Cape Cod was declared a sole-source aquifer by EPA in 1982. Unconfined conditions, highly permeable, sandy soils, and shallow depths to ground water make this aquifer system highly susceptible to contamination from various activities related to the rapid growth and development of the area. Furthermore, all of the Cape is surrounded by sea water, making it vulnerable to salt water intrusion.

Ground Water Quality Issues

Massachusetts

In general, Massachusetts has many of the same problems and concerns related to ground water quality that affect other northeastern states. The greatest threats to water quality in the stratified drift aquifer of Cape Cod are related to incompatible land use activities, which include sewage disposal through private septic systems and municipal waste water treatment plants, landfills, dumps, road salt, application of agricultural chemicals, surface impoundments, leaking underground storage tanks, accidental chemical spills, industrial discharges, and toxic household chemicals. In some areas, public and private wells have been adversely affected by sea water intrusion.

Cape Cod

The Cape Cod area has experienced significant ground water contamination problems from leaking underground motor-fuel tanks, sewage discharges, and salinity increases from road salt and sea water intrusion. Underlying much of the concern on Cape Cod is its rapid population growth and development; the population increased by more than 50 percent during the 1970s, and ground water management efforts are hard put to keep pace with residential and commercial development pressures.

TABLE 3.3 Aquifer and Well Characteristics in Massachusetts

Aquifer name and description	Well characteristics				Remarks
	Depth (ft)		Yield (gal/min)		
	Common range	May exceed	Common range	May exceed	
Stratified-drift aquifer:[1] Sand and gravel with silt, glacial outwash, ice-contact, and delta deposits; some beach and dune deposits included. Moraines also contain till. Generally unconfined, locally confined.	60 – 120	200	100 – 1,000	2,000	Used extensively for public supply; also used for industry, fish hatcheries, agriculture, and rural supplies. Locally, large iron or manganese concentrations a problem. Some saline water intrusion in coastal areas. Low pH of water may corrode pipes and appliances.
Sedimentary bedrock aquifer: Red sandstone, shale, arkosic conglomerate, and basaltic lava flow. Generally, unconfined, confined at depth.	100 – 250	500	10 – 100	500	Used for rural supplies and some industry. Deep wells produce hard water.
Carbonate rock aquifer: Limestone, dolomite, and marble. Confined.	100 – 300	1,000	1 – 50	1,000	Used for rural supplies and some industry Water hard.
Crystalline bedrock aquifer: Metamorphic and igneous rock predominantly gneiss and schist. Confined.	100 – 400	1,000	1 – 20	300	Used for rural supplies. Locally, large iron concentrations a problem. Recently drilled wells generally deeper than older wells. Low pH of water may corrode pipes and appliances.

[1]Well depths and yields reported for stratified drift are for public supply wells. Rural domestic wells yield 5 to 80 gal/min from 1½ to 2½ inch diameter well screens, 3 to 5 ft in length.

SOURCE: U.S. Geological Survey, 1984.

Ground Water Management and Protection

Massachusetts

The Commonwealth of Massachusetts has an intergovernmental ground water protection strategy and program involving state, county, and municipal governments. The Massachusetts Department of Environmental Quality Engineering (DEQE) has regulations to control ground water discharges from industrial sources and all sources discharging more than 15,000 gallons per day of sanitary wastes. DEQE also has comprehensive regulations to control hazardous and solid wastes, land application of sludge and septage, and underground injection of wastes. The DEQE also has a strong technical and financial assistance program for local government.

The technical assistance program includes a bimonthly newsletter and a number of guides and handbooks on best management practices for road salting, erosion and sedimentation, developing a ground water protection program, ground water monitoring, and hydrogeologic information sources for the state. The DEQE has also developed a unique water supply protection atlas consisting of four overlays: water sources, waste sources, aquifer information, and drainage basins for 177 of the USGS topographic quadrangle maps that cover the state. Each municipality has been given overlays for the 125,000 topographic maps delineating its boundaries. This has been an important tool in developing state and local ground water protection programs. DEQE has recently hired a full-time land use planner to assist municipalities in development and implementation of local land use controls to protect ground water.

Massachusetts also has a financial assistance program to municipalities for acquisition of land to protect important aquifers. This program provided $10 million in grants to 26 communities in 1983 and 1984 with another $4 million available for 1986. The Massachusetts legislature failed to pass a major piece of legislation in 1985 that would have allowed local communities to tax real estate transactions and create a fund for the purchase of land for a number of purposes including aquifer protection. However, this legislation has now been refiled.

A number of cities and towns in Massachusetts have developed programs to protect ground water. At present, 325 of the 351 cities and towns have adopted zoning regulations, by-laws, or Board of Health regulations to protect ground water. These include flood plain and wetland zoning, aquifer protection districts, toxic and hazardous materials controls, and control of underground storage of petroleum products.

Cape Cod

Over the past decade, Cape Cod has developed a unique legislative local and regional ground water management and regulatory program involving the cooperative effort of townships, county government, and state and federal agencies. The effort for ground water protection on Cape Cod was started under the 208 program of the federal Clean Water Act in 1975. Cape Cod was given a grant to develop the Water Quality Management Plan/ environmental impact statement for Cape Cod through EPA. The program developed by the Cape Cod Planning and Economic Development Commission (CCPEDC) was very successful and became recognized as a national prototype. It has obtained additional federal and state funds, and has received considerable press coverage. There was extensive public participation and input throughout the planning process, and this is still the case. This effort is all the more impressive because CCPEDC's role is purely advisory, a feature that may be responsible for the program's success. The program has also involved extensive participation by the USGS, as well as cooperation from the National Park Service, the Massachusetts Department of Environmental Quality and Engineering, the Barnstable County Health and Environmental Department, and local communities.

The assistance of the USGS has enabled Cape Cod to define more clearly sensitive areas and has facilitated closer scrutiny of existing land use controls. One of the most important outputs of the 208 program has been the development and implementation of three model health regulations or by-laws adopted by many local communities: one dealing with water resource protection district zoning overlays; another addressing underground storage tank regulation; and a third concerning the storage, use, and disposal of toxic and hazardous materials. These three health regulations or by-laws have been adopted at the local level by a majority of the 15 towns on Cape Cod. Local actions to protect ground water go well beyond adoption of by-laws and regulations. Many of the communities have developed protective mechanisms such as transfer of development rights, granting of conservation easements, performance standards for development, open space preservation, and public education.

A second major highlight of the Cape Cod program is the development of a nitrogen-loading formula, which has been used extensively to predict the probable concentration of nitrate/nitrogen in public supply wells attributable to residential development. A third accomplishment is the delineation of recharge zones of contribution for all 140 public supply wells on Cape Cod.

The CCPEDC has recently become involved in a program with the EPA, the Massachusetts DEQE, and the USGS that integrates the local and regional interests with the state and federal activities. The program, called the Cape Cod Aquifer Management Project, seeks to coordinate several different levels of government concerned with ground water management. By focusing on a target aquifer, the participating agencies will attempt effective allocation of resources to achieve concrete environmental results. One of the most important conclusions reached by CCPEDC is that ground water protection cannot be done exclusively at one level of government and that use of expertise at different levels is necessary to achieve meaningful results.

The existing program of CCPEDC has been successful in reducing problems or threats related to underground storage tanks, small generators (including households) of toxic and hazardous wastes, and high-density residential development. The program relies heavily on local action, frequently in the form of ordinances and zoning restrictions, which in turn require intercommunity cooperation, public awareness, and citizen involvement.

COLORADO

Overview of Ground Water Resources

Ground water constitutes 18 percent of the total water used in Colorado and, in some areas, is the main source for domestic and irrigation supply. Approximately 15 percent of the total population gets its drinking water supply from ground water. Ground water withdrawals for irrigation are 96 percent of total ground water withdrawals.

Geographic and topographic features cause significant differences in ground water availability and conditions from one part of the state to another. Major areas based on geology, topography, drainage, and physiography are the South Platte River basin, the Arkansas River basin, and the High Plains in eastern Colorado; the Rocky Mountain area in central Colorado; and western Colorado Plateau country.

Colorado has seven principal aquifers or aquifer systems. Four of the principal aquifers consist of unconsolidated deposits and include the alluvial aquifer along the South Platte River and its tributaries, the High Plains aquifer underlying the High Plains, and the San Luis Valley aquifer system in the Rocky Mountains area. Most ground water withdrawal, which in Colorado is for irrigation, comes from the aquifers in the unconsolidated deposits. The remaining three aquifers consist of consolidated rock and include the Denver Basin aquifer system, the Piceance Basin aquifer system in western Colorado, and the Leadville limestone aquifer in the Rocky Mountain area (see Table 3.4).

TABLE 3.4 Aquifer and Well Characteristics in Colorado

Aquifer name and description	Well characteristics: Depth (ft) Common range	Depth (ft) May exceed	Yield (gal/min) Common range	Yield (gal/min) May exceed	Remarks
Principal Aquifers					
Unconsolidated sedimentary rock aquifers:					
South Platte alluvial aquifer: Interbedded gravel, sand, silt, and clay; contains some cobbles and boulders; unconsolidated. Generally unconfined.	30 - 150	250	100 - 1,500	3,000	Provides water for public supplies and supplemental irrigation. Transmissivity ranges from 2,000 to 200,000 ft^2/d. Dissolved-solids concentration ranges from 100 mg/L in areas overlain by dune sand to about 4,000 mg/L in some downstream areas. Water hard to extremely hard. Local areas show significant water-level declines.
Arkansas alluvial aquifer: Boulders, cobbles, gravel, sand, and clay. Generally grades from fine sand near the surface to coarse sand and gravel at the base. Generally unconfined.	25 - 100	200	100 - 1,200	1,500	Principal source of water for irrigation, public supply, and industrial wells. Transmissivity ranges from 1,000 to 150,000 ft^2/d. Dissolved-solids concentration ranges from about 800 to 5,000 mg/L. Water hard to extremely hard.
High Plains aquifer: Gravel, sand, silt, and clay; contains some caliche. Poorly to moderately consolidated. Generally unconfined.	200 - 400	450	350 - 2,000	2,500	Primary source for irrigation, public supply, and domestic use. Transmissivity ranges from 3,000 to 30,000 ft^2/d. Dissolved-solids concentration generally ranges from 200 to 500 mg/L. Widespread water-level declines affecting well production and increasing irrigation costs.

San Luis Valley aquifer system:					
Unconfined aquifer: Clay, silt, sand, and gravel; unconsolidated. Alluvial and lacustrine. 0 to 200 ft thick.	50 – 150	150	500 – 1,200	2,000	Provides supplemental irrigation water. Withdrawals greatest in Rio Grande and western Alamosa Counties. Transmissivity ranges from 100 to 34,000 ft^2/d. Dissolved-solids concentration ranges from 72 to 31,200 mg/L. Local areas show water-level declines.
Confined aquifer: Clay, silt, sand, and gravel, unconsolidated, interbedded with lava flows and tuffs. As much as 19,000 ft thick.	300 – 800	2,000	500 – 1,200	2,000	Provides supplemental irrigation water. Withdrawals greatest in Conejos and western Saguache Counties. Transmissivity ranges from 200 to 200,000 ft^2/d. Dissolved-solids concentration ranges from 60 to 2,440 mg/L.
Consolidated sedimentary rock aquifers:					
Denver Basin aquifer system:					
Dawson aquifer: sandstone and conglomerate with interbedded shale, siltstone. Confined except near outcrop area.	200 – 1,000	1,400	5 – 150	300	Sandstone thickness ranges from 100 to 400 ft. Dawson is uppermost aquifer in group. Primarily used for rural and public supply. Potential for local contamination from Lowry landfill in Arapahoe County. Less than 200 mg/L dissolved solids.
Denver aquifer: Sandstone with interbedded shale, siltstone, and coal. Confined except near outcrop area.	200 – 1,500	2,100	5 – 100	300	Sandstone thickness ranges from 100 to 300 ft. Denver contains more shale than other aquifers in group. Used primarily for domestic supply. Generally less than 200 mg/L dissolved solids.
Arapahoe aquifer: Sandstone and conglomerate with interbedded shale, siltstone. Confined except near outcrop area.	200 – 2,000	2,600	10 – 600	800	Sandstone thickness ranges from 100 to 350 ft. Arapahoe most permeable aquifer in group. Used extensively for public, commercial, and domestic supply. Less than 500 mg/L dissolved solids.

SOURCE: U.S. Geological Survey, 1984.

Ground Water Quality Issues

Major sources of contamination of ground water resources in Colorado include return flows from agricultural use, illicit dumping, and septic tanks. Other significant sources are municipal waste discharges, inadequate facilities for toxic waste disposal, and mining activities. Minor sources are sanitary landfills and leaking underground storage tanks.

Other new problems are developing within the state and are being handled on a daily basis, such as backflow of agrichemicals down water wells, contamination from exploratory wells from drilling activities, land application of treated waste water, and industrial impoundments with a discrete discharge point to surface waters.

Ground Water Management and Protection

Ground water rights administration was first legislated in 1957. The major ground water legislation came in 1965 with the Ground Water Management Act. This legislation allowed designation of basins and formation of ground water management districts. A 1969 amendment required augmentation for certain ground water rights.

In 1981, the governor of Colorado gave the Colorado Department of Health (CDH) a charge to review all the environmental laws of the state to see how they protected ground water and to coordinate 14 agencies in the state to develop a ground water protection plan. The CDH developed the protection plan and had all 14 agencies sign a memorandum of agreement in 1985. The agreement states that the agencies will enforce their laws and that if they encounter ground water problems or pollution activities they will use the ground water strategy developed by the CDH.

The CDH strategy includes a classification system that is similar to Florida's system. The initial classification is based on beneficial uses; the second part includes numeric standards; then a screening mechanism is employed for site-specific areas (hydrogeologic studies); and the next step is to write specific control regulations for activities that are not regulated for ground water impacts. The CDH also has an advisory committee that includes representation from the Colorado Association of Commerce and Industry, the League of Women Voters, and the Bar Association.

CONNECTICUT

Overview of Ground Water Resources

Ground water constitutes an important natural resource that supplies domestic water to approximately one third of the state's 3.1 million people. Its degree of importance is increasing because of limited land areas available for surface reservoirs and the cost of developing additional surface systems. According to the USGS *National Water Summary 1984*, approximately 11 percent of the total fresh withdrawals within the state were from ground water. Approximately 17 percent of public water supplies are from ground water; essentially all of rural domestic supplies are from ground water.

Connecticut has a variety of physiographic and geologic conditions that cause considerable variability in ground water conditions throughout the state. Two principal types of aquifers underlie Connecticut—unconsolidated stratified glacial drift aquifers, which are composed of sand and gravel, and bedrock aquifers composed of sedimentary, igneous, and metamorphic rocks. The stratified drift aquifers are the most productive sources of ground water in the state. These glacially derived sedimentary aquifers occur mainly in the river valleys of the state and are particularly predominant in the Connecticut Valley lowland.

Bedrock aquifers underlie the entire state. They are the principal source of water for rural homes and small public supplies, as well as some commercial establishments and industries. These aquifers can be subdivided into a sedimentary system, which is composed of sandstone, shales, siltstones; the igneous crystalline system, which is composed predominantly of granite, metamorphic gneiss, and schist; and the carbonate system, which is composed of marble. The sedimentary system underlies the Connecticut Valley lowland, with the carbonate system occupying a small area in the western part of Connecticut. The crystalline aquifer underlies the eastern and western parts of the state. Well yields in both of the sedimentary and crystalline bedrock systems are generally low but sufficient to supply small demands. Water quality in the bedrock aquifers is generally suitable for most uses.

The largest ground water withdrawals are concentrated in Hartford, Fairfield, Middlesex, and New Haven counties, where most of the water is used for public supply. The majority of this withdrawal is from the stratified drift aquifers (see Table 3.5).

Ground Water Quality Issues

The stratified drift aquifers are the most susceptible to contamination and are also the most important ground water resources of the state. The

TABLE 3.5 Aquifer and Well Characteristics in Connecticut

Aquifer name and description	Well characteristics: Depth (ft) Common range	Depth (ft) May exceed	Yield (gal/min) Common range	Yield (gal/min) May exceed	Remarks
Stratified-drift aquifers: Sand and gravel, commonly with interbedded layers or lenses of silt and clay. Generally unconfined.	50 – 100	150	50 – 500	2,000	Largest yields from wells near major rivers. Iron and manganese concentrations commonly exceed 0.3 and 0.05 mg/L, respectively. Dissolved-solids concentrations range from 31 to 1,270 mg/L. Salty ground water present locally in coastal areas. Aquifers susceptible to contamination.
Sedimentary-aquifer system: Sandstone, shale, siltstone, and conglomerate; some interbedded basalt flows and dikes. Unconfined to partly confined in upper 200 ft, may be confined at depth.	100 – 300	500	2 – 50	500	Hydrologic characteristics poorly defined, particularly in zones deeper than 300 ft. Generally overlain by variable thicknesses of unconsolidated deposits. Moderately hard to hard water, and large concentrations of dissolved chloride sodium, and sulfate occur locally.

Crystalline bedrock aquifer (noncarbonate rocks): Gneiss and schist with minor amounts of other metamorphic and igneous rock types. Generally unconfined in upper 200 ft, may be confined at depth.	100 - 300	500	1 - 25	200	Hydrologic characteristics poorly defined, particularly in zones deeper than 300 ft. Generally overlain by variable thicknesses of unconsolidated deposits. Iron and manganese concentrations may exceed 0.3 and 0.05 mg/L, respectively. Dissolved-solids concentrations range from 20 to 1,590 mg/L.
Carbonate rock aquifer: Marble; some schist and quartzite zones. Generally unconfined in upper 200 ft, may be confined at depth.	100 - 300	500	1 - 50	200	Hydrologic characteristics poorly defined, particularly in zones deeper than 300 ft. Generally overlain by variable thicknesses of unconsolidated deposits. Generally hard to very hard water; large iron and manganese concentrations are local problems.

SOURCE: U.S. Geological Survey, 1984.

major sources of ground water contamination include waste disposal facilities, underground fuel and chemical storage tanks, accidental spills, and pesticide and road salt storage and application. Additional ground water quality problems have developed historically in areas where the stratified drift aquifers are adjacent to salt water bodies; excessive pumping and coastal flooding have caused salt contamination of some of these aquifers.

Ground Water Management and Protection

Connecticut has been a national pacesetter in statewide programs for ground water protection. It was one of the first states to adopt and implement a comprehensive ground water classification system intergrated with water quality standards, land use policies, and discharge permits. The backbone of the Connecticut system is a four-class ground water classification system. The entire state has been mapped and classified according to these four classes of ground water. The most protected class applies to water utility and municipal drinking water supplies. The next two classes apply to private drinking water supplies and water supplies that may not be suitable for potable use unless treated because of existing or past impacts on the water quality. The final class designates areas in which certain treated industrial waste water and major residential waste disposal practices are allowed and there are no future plans to use the ground water as a source of drinking water. The Connecticut ground water protection system recognizes the intimate connection of surface waters and ground waters and manages them in a systems approach. Although the Connecticut system is not without problems and controversy, it appears to be an effective system for managing and controlling activities that have an impact on ground water quality. For instance, it has been used on occasion to close landfills and prevent location and siting of other activities, industries, or operations that could potentially have an impact on ground water quality. One important aspect of the program is an active effort to work closely with industries and local governments to plan and implement strategies for location and permitting of certain activities that might have an adverse impact on ground water. It is administered by the Connecticut Department of Environmental Protection (DEP).

The Connecticut system was adopted in 1980 and was implemented successfully, partly because an extensive data base existed on the hydrogeologic conditions of the state, which has been developed through the state's cooperative program with USGS. Over a period of several years, the Connecticut DEP has collected information on all surface watersheds, the properties and distribution of aquifers, depth of water tables, water quality of sensitive lands and water courses, locations of all existing public water supply wells, locations of waste water discharges, industrial waste lagoons, landfills, oil and chemical spills and leaks, salt piles, agricultural waste disposal, septic tank distributions, and a variety of other information needed in a comprehensive water quality management program. Connecticut's system is based primarily on water use criteria rather than on discharge criteria. The basic policy of the state program is to restore or maintain the quality of ground water to a quality consistent with its use for drinking without treatment. The program includes corrective actions as well as restrictive actions to maintain or improve ground water quality up to the next highest class at least. One of the problems with the program is the shortage of the areas classified for waste disposal activities. There, of course, is also public resistance to the classification of areas for waste disposal or landfill activities. This has resulted in a potential shortage of areas suitable for landfills and other critical activities needed in the state and is one of the factors encouraging development of resource recovery facilities. The success of the Connecticut program is due partly to an active and intimate public education and involvement program, which includes direct assistance to planning, zoning, and conservation commissions in land and water use planning.

FLORIDA AND DADE COUNTY

Overview of Ground Water Resources

Ground water is an abundant and vital resource in Florida. Large quantities of water are obtained from each of the principal aquifers in most areas of the state. Because of its abundance and availability, ground water is the principal source of fresh water for public supply and rural and industrial uses and is the source of about half of the water used for irrigation. About 90 percent of Florida's population depends on ground water for its drinking water, according to the USGS *National Water Summary 1984*. Among other states, Florida ranks very high in its use of ground water: eighth in total fresh water withdrawal; first for rural, domestic, and livestock; second for public supply; third for industrial uses; and ninth for irrigation withdrawals. Florida's exceptionally abundant ground water resources are the result of a combination of favorable climatic, physiographic, and geologic conditions. The state is largely composed of flat-line relatively permeable carbonate rocks overlain by very permeable sedimentary sands. Its low relief and relatively high rainfall of more than 50 inches per year contribute to high recharge rates to the enormous ground water reservoir provided by the surface sediments and carbonate rock.

Ground water is especially important in Dade County in southeast Florida, which contains Miami and surrounding communities. In this area the principal aquifer is the shallow Biscayne aquifer, which is composed of highly permeable limestone that is due to extensive carbonate dissolution. The Biscayne aquifer, particularly in the Dade County area, is very vulnerable to contamination and is the sole source of drinking water for more than 3 million people in this part of Florida.

Another major aquifer in Florida is the sand and gravel aquifer that is a major source of ground water in the western part of the Florida panhandle. As the name implies, this aquifer is composed of mixed surficial sediments. It serves as a major water supply for Pensacola and associated industries near Pensacola and is vulnerable to contamination by many of these industries.

The southern third of Florida contains a series of unnamed surficial and intermediate aquifers that are a major source of ground water. These aquifers are composed of surficial sediments and underlying carbonate beds and interbedded sand, silts, and clays. These aquifers are important public supplies for some communities and are widely used for rural supplies. The surficial aquifers are highly vulnerable to contamination from agricultural activities and other sources, while the intermediate aquifers are less vulnerable to contamination from source activities. Locally, these aquifers may be subject to contamination from saline water intrusion and elevated natural concentrations of radon-226.

The most extensive and productive aquifer system in Florida is the Floridan aquifer, which extends across the entire state, southern Georgia, and adjoining small parts of Alabama and South Carolina. The Floridan aquifer consists of as much as 3500 feet of limestone and dolomite beds, which occur at or near the land surface in the western part of the peninsula. In other parts of the state this aquifer is buried to depths as much as 1500 feet below sea level, making it a confined aquifer. The Floridan aquifer serves as a public water supply for many large cities and communities such as Jacksonville, Orlando, St. Petersburg, and Tallahassee and is also used as a major source of irrigation in rural areas. Where the aquifer is at or near the surface, it is susceptible to contamination by agricultural activities, leaching from landfills, and other waste disposal activities. Contamination in surficial and intermediate aquifers by pesticides from agricultural applications in some parts of the state has received considerable attention recently. Nonpotable confined parts of this aquifer are locally used for injection of industrial and municipal waste waters in some parts of the state (see Table 3.6).

Ground Water Quality Issues

Because of Florida's heavy dependence on ground water and the vulnerability of this resource to contamination from a variety of sources, there are a number of significant concerns relating to ground water contamination and protection. The following sources of contamination are considered to be of principal concern. More than 6000 pits, ponds, and lagoons for industrial sources of waste waters have been identified within the state. Agricultural activities are contributing significantly to contamination of ground water by pesticides and fertilizers. This has been particularly highlighted by the contamination of ground water by aldicarb and ethylene dibromide in central Florida, which received widespread national publicity. Another major area of concern is that of leaking underground storage tanks for fuels and chemicals. More than 40,000 underground storage tanks have been identified within this state, and considerable attention is being paid to protecting ground water from possible leakage of the chemicals and fluids in these tanks. Salt water intrusion is a major natural source of ground water degradation in many areas where pumping draws in salt water or declining water tables and other factors effect salt water intrusion into potable supplies. Florida ranks among the top five states in the nation for the number of uncontrolled hazardous waste sites identified within the state. Landfills, mining activities, and a variety of industrial sources round out the list of major sources of contamination receiving attention in Florida.

Ground Water Management and Protection

Florida has established a comprehensive program of policy, regulations, and governmental framework to manage its ground water resources both in quality and in quantity. Ground water quality is managed primarily through the Department of Environmental Regulation at the state level. The ground water quality program is carried out through six district offices of the Florida Department of Environmental Regulation. There are five water management districts that overlay the entire state, and their concern is the management of the water resources within their district. To that extent, they become involved in the quantity of ground water pumpage and recharge. The department carries out ground water quality regulations and protection activities. Whether an activity relates to ground water quality carried out by the department's own district offices or water supply carried out by one of the five water management districts, activities are tailored to the specific area of the state on a site-specific basis. For the department to be most effective, contracts have been let with the water management districts for the ambient monitoring of some wells. Other coordination exists where efficiency can be maximized by such cooperation.

Florida has one of the most thorough and active ground water protection programs in the country. Its program is based on a combination of several approaches that include a ground water classification system, permitting

TABLE 3.6 Aquifer and Well Characteristics in Florida

Aquifer name and description	Water withdrawals (Mgal/d)	Well characteristics: Depth (ft) Common range	Well characteristics: Yield (gal/min) Common range	Well characteristics: Yield (gal/min) May exceed	Remarks
Surficial aquifers:					
Biscayne aquifer: Limestone, sandstone, and sand. Unconfined.	461	40 - 150	500 - 1,000	7,000	Supplies all public-supply water systems in southern Palm Beach, Broward, and Dade Counties. Designated by U.S. Environmental Protection Agency as "sole-source" drinking-water supply. Aquifer managed carefully to control saltwater intrusion into coastal well fields. Water generally very hard.
Sand-and-gravel aquifer: Sand and gravel interbedded with discontinuous clay layers. Unconfined in upper part to locally confined in deeper part.	34	100 - 300	500 - 1,000	2,000	Primary water source for Pensacola and other public-supply and private pumpage in Escambia and Santa Rosa Counties. Water soft; little dissolved solids (less than 50 mg/L), but locally iron exceeds 0.3 mg/L. Known as Pliocene–Miocene aquifer in Alabama.
Unnamed surficial aquifers: Sand, shell, and clayey sand; locally contains thin discontinuous limestone layers. Unconfined to locally confined.	104	50 - 400	<100	1,000	Locally important as water sources where deeper aquifers contain saline water, especially along east coast and in southwest Florida. Hardness and dissolved-solids concentrations vary widely. Saltwater intrusion a local problem.

Intermediate aquifer(s): Limestone and shell beds with discontinuous clay layers and some interbedded sand. Confined.		100 - 600	<200	1,000	Important public-supply source along west coast from Sarasota to Lee County. Elsewhere tapped generally for small to moderate supplies. Flowing wells common in coastal areas. Some parts in and around Sarasota County yield water containing sulfate and radionuclide concentrations exceeding National drinking-water regulations. Also called "secondary artesian aquifer(s)."
Floridan aquifer system: Limestone and dolomite. Unconfined in outcrop areas, confined where deeply buried.	460	100 - 1,800	500 - 1,000	20,000	Occurs throughout Florida and extends into parts of Alabama, Georgia, and South Carolina. Contains nonpotable, saline water in south Florida, westernmost Florida panhandle, and locally along the west coast where unconfined. Elsewhere water is hard. Locally sulfate concentrations exceed National drinking-water regulations. Principal source of water for all uses where water is fresh. Also called "principal artesian aquifer" and "Floridan aquifer."

SOURCE: U.S. Geological Survey, 1984.

systems for waste discharges and other surface activities, ground water quality standards, an underground injection control program, a statewide ground water quality monitoring program, various public awareness and involvement programs, and research programs on the behavior and fate of specific organic contaminants in ground water systems.

The ground water classification system consists of four classifications ranging from sole-source drinking water supplies in which essentially no degradation is allowed to confined nonpotable saline aquifers in which degradation activities such as underground injection of waste effluents can be permitted. The state has been aggressive in setting ground water standards for specific compounds that include the normal priority pollutants as well as additional organic compounds for which EPA has not yet set standards. In addition, the state regulations allow state officials to set standards on a case-by-case basis if needed for any compound. The state ground water quality monitoring network is aimed at assuring high-quality drinking water, general compliance to state standards and regulations, monitoring of the condition of known contaminated areas, and early detection of additional problems in ground water quality.

One of the unique aspects of the Florida approach to ground water protection is the special effort to protect the zones of influence around public supply wells. This system is based on the time that it takes for a contaminant to travel from its source to the well. The shorter the time of travel in ground water toward the production well, the less protected is that area. This is based on land use controls in which certain activities and facilities that may be sources of contamination, such as underground storage tanks or landfills, and other activities might be prohibited from designated zones near the production well.

In addition to state and district regulations, several counties and communities or other jurisdictions have developed more specific and restrictive ground water protection regulations. Dade County in southeast Florida represents a good example of such a jurisdiction. Because of its high population and dependence on the highly vulnerable Biscayne aquifer, Dade County has established an aggressive approach to strict protective measures. This is based largely on land use controls but also on a number of other mechanisms. Dade County, for instance, was the pioneer in developing the concept of protecting the zone of influence around production wells from activities that could be potentially contaminating. These measures are based largely on zoning and land use restrictions that can prohibit underground storage tanks and other potentially contaminating activities near wells. Dade County works cooperatively with two other counties in a very strong regional program entitled "The Biscayne Aquifer Project," which was established to protect that vital and important aquifer. Dade County has established a broad definition of what constitutes hazardous wastes and includes a list of more than 900 chemicals, which is much broader than any state or federal EPA regulation. It has also identified 8000 generators of hazardous materials and closely regulates the activities of those generators through permits. In addition to land use and zoning controls, the county has been successful in incorporating other tools such as recharge area management, growth management, monitoring, and conventional pollution control techniques. The county was also a pioneer in developing comprehensive and strict underground storage tank control regulations. These regulations are setting the pace for many other state regulations on underground storage tanks.

KANSAS

Overview of Ground Water Resources

Kansas relies on ground water resources for municipal, rural, industrial, and irrigation water supplies. The climate ranges from humid areas in the east to semiarid areas in the west. Annually, Kansas exports about 10 million acre-feet of surface water in addition to having about 400 million acre-feet of ground water in storage.

Ground water supplies about 5.6 billion gallons per day or 85 percent of the water used in Kansas. Municipal and rural systems provide ground water to almost 1.2 million people (approximately 49 percent of the state's population). Most of the ground water withdrawn is used for irrigation (about 93 percent).

Ground water conditions differ with physiography and geology. Physiographic provinces in Kansas are the Osage Plains and Dissected Till Plains sections of the Central Lowlands province, the Ozark Plateaus province, and the Great Plains province (Fenneman, 1946).

Principal aquifers in Kansas consist of two types—unconsolidated gravel, sand, silt, and clay and consolidated sandstone, limestone, and dolomite. The principal aquifers are the Alluvial aquifers, Glacial-Drift aquifer, High Plains aquifer, Great Plains aquifer, Chase and Council Grove aquifer, Douglas aquifer, and Ozark aquifer. Table 3.7 indicates these aquifer and well characteristics.

Ground Water Quality Issues

Kansas has had a long legacy of pollution from the oil and gas industry. In addition, about half a million wells have been either drilled or abandoned and not adequately plugged, and these have the potential to cause problems

TABLE 3.7 Aquifer and Well Characteristics in Kansas

Aquifer name and description	Well characteristics			Remarks
	Depth (ft)	Yield (gal/min)		
	Common range	Common range	May exceed	
Alluvial aquifers: Quaternary fluvial deposits of clay, silt, sand, and gravel. Generally unconfined.	10 – 150	10 – 500	1,000	Well yields in Kansas, Arkansas, Republican, and Pawnee River valleys exceed 500 gal/min. Wells in other valleys usually yield less than 100 gal/min. Locally, water from alluvial aquifers can have large concentrations of dissolved solids, chloride, sulfate, nitrate, iron, and manganese. Large concentrations of selenium and naturally occurring gross-alpha radioactivity sometimes occur in water from northern part of Great Plains.
Glacial-drift aquifer: Pleistocene glacial deposits of clay, silt, sand, and gravel. Generally unconfined.	10 – 300	10 – 100	500	Water from shallow wells generally a calcium bicarbonate type with less than 500 mg/L dissolved solids, but large concentrations of nitrate can occur. Water from deep wells can have large concentrations of dissolved solids, chloride, sulfate, iron, or manganese.
High Plains aquifer: Fluvial and eolian deposits of clay, silt, sand, and gravel of Cenozoic age. Generally unconfined.	10 – 450	500 – 1,000	1,500	Water generally a calcium bicarbonate type with concentrations of dissolved solids less than 500 mg/L, but large concentrations of fluoride and selenium can occur in northern Great Plains. Provides water supplies for Dodge City, Garden City, Great Bend, Pratt, Hutchinson, McPherson, Wichita, and most other towns in Great Plains.

Great Plains aquifer: Dakota and Cheyenne Sandstones of Cretaceous age. Generally unconfined.	20 - 200	10 - 100	1,000	Water quality variable. Calcium bicarbonate type water with less than 500 mg/L of dissolved solids produced where the aquifer is exposed. Sodium bicarbonate or sodium chloride type water with large concentrations of dissolved solids is produced west and north of the surface exposure. Large concentrations of iron occur in water from some wells. Some wells in Finney, Ford, and Hodgeman Counties can yield more than 1,000 gal/min.
Chase and Council Grove aquifer: Limestones of Chase and Council Grove Groups of Permian age. Generally unconfined.	20 - 200	10 - 20	200	Water generally a calcium bicarbonate type with concentrations of dissolved solids less than 500 mg/L. Water from some wells can have large concentrations of sulfate. Wells in Butler and Cowley Counties can produce water with large concentrations of dissolved solids. Concentrations of dissolved solids and chloride large west of the surface exposure, and water is not used.
Douglas aquifer: Channel sandstone of Pennsylvanian age. Generally unconfined.	5 - 400	10 - 40	100	Water ranges from a calcium bicarbonate type, with less than 500 mg/L of dissolved solids where aquifer is exposed, to a sodium bicarbonate or sodium chloride type, with large concentrations of dissolved solids at depth or west of surface exposure. Concentrations of fluoride may be large. Equivalent to Vamoosa-Ada aquifer in Oklahoma.
Ozark aquifer: Weathered and sandy dolomites of Arbuckle Group. Cambrian and Ordovician age. Confined.	500 - 1,800	30 - 150	500	Water generally a calcium bicarbonate type with less than 500 mg/L of dissolved solids in the Ozark Plateaus and in extreme southeast corner of the Osage Plains. Sodium bicarbonate chloride or sodium chloride type water with large concentrations of dissolved solids is produced in rest of Osage Plains. Hydrogen sulfide gas, or large concentrations of gross-alpha radioactivity or iron, can occur in water from some wells. Equivalent to Roubidoux aquifer in Oklahoma.

SOURCE: U.S. Geological Survey, 1984.

in terms of ground water pollution. About 11 million tons of salt per year are brought to the surface by the oil and gas industry. Other major sources of ground water contamination that cause problems in this state are agricultural nonpoint drainage, storage or disposal of either solid or liquid waste or hazardous materials, mineralized water intrusion, accidental spills, landfills, underground storage tanks, underground pipelines, injection/enhanced recovery wells, industrial disposal wells, chemigation (chemicals added to irrigation water), and mineral mining.

Ground Water Management and Protection

Kansas has long supported programs for the protection of the environment. Specific legislation for the protection of ground water was first enacted in 1931. The state has five agencies administering laws spread over five chapters of the statutes with an annual expenditure of $12 million and a staff of more than 350 persons involved in water regulation related to quality.

The regulation of wells that penetrate into or through ground water is handled by three different agencies. Oil and gas regulation is administered by the Kansas Corporation Commission, which has a mandate to protect fresh ground water supplies from adverse effects of mineral development activities. The Kansas State Board of Agriculture issues permits for water withdrawals and administers laws related to conservation and use of water resources, including appropriation of ground water; it also assists with the organization of Ground Water Management Districts. The Kansas Department of Health and Environment has regulatory authority over matters dealing with public water supplies and water pollution. It is also responsible for collecting, analyzing, and interpreting ground water quality data; developing water quality management plans; and responding to emergency water pollution problems.

The main elements of a ground water protection strategy are found in the state's public health, environmental, and conservation laws. Parts of the strategy are expanded upon in the Kansas Water Plan, the Water Quality Management Plan, and the Ground Water Quality Management Plan.

Kansas also has Ground Water Management Districts that are locally managed political subdivisions of the state and were formed as a result of the Ground Water Management District Act of 1972. There are currently five of these districts: (1) western Kansas, (2) Equus beds, (3) southwest Kansas, (4) northwest Kansas, and (5) Big Bend. Each district is charged with managing ground water resources within its boundaries.

The involvement of ground water management districts represents a careful effort to get more local involvement in decision-making and has resulted in decisions beneficial to the protection of ground water. For example, they document the need and recommend measures to contain the movement of contaminants under the intensive ground water management strategies. Control strategies for the protection of ground water include licensing of well drillers and permitting of gas, oil, and water wells. The laws provide for the conservation of gas and oil as well as protection of water aquifers. The state's solid waste, toxic waste, and clean water laws regulate generators, transporters, and disposers of hazardous or toxic waste. The zero pollution strategy and the consistency of broad-based support for the program are among the program's most important features.

NEW YORK

Overview of Ground Water Resources

Six million of New York's 17.5 million residents rely on ground water for drinking supplies. Half of these live on Long Island, where ground water withdrawals for all uses total 486 million gallons per day. Upstate New York, which consists of the boroughs of all counties north of Bronx, Manhattan, and Richmond (Staten Island), is discussed in this section.

Primary and principal aquifers in upstate New York consist of unconsolidated glacial stratified-drift and valley-fill deposits and consolidated clastic and carbonate sedimentary rocks, some of which have been metamorphosed. Characteristics of such aquifers are also described in Tables 3.8 and 3.9. Most portions of upstate New York are characterized by hilly or mountainous terrain, intersected by numerous valleys. The slopes and hilltops in this region are generally covered by relatively dense glacial till, bedrock is relatively near the land surface, and the amount of ground water that can be withdrawn is limited. Low-yielding "nonaquifer" areas make up about 90 percent of the total upstate land area. However, about 2 million upstate residents do rely on ground water in these areas for individual household supplies.

In contrast, highly permeable, high-yielding sand and gravel deposits are often found in the valley areas. The New York State Department of Environmental Conservation (DEC) has adopted the following terminology to identify these aquifers:

- "Primary" water supply aquifers are high-yielding aquifers that currently serve sizable municipal populations as their source of municipal supply. These underlie 18 discrete areas with populations ranging from 8,000 to 150,000. A total of roughly 700,000 people rely on municipal supplies in these areas.

TABLE 3.8 Aquifer and Well Characteristics in New York

Aquifer name and description	Well characteristics			Remarks
	Depth (ft)	Yield (gal/min)		
	Common range	Common range	May exceed	
Upstate				
Stratified-drift–Lacustrine and ice-contact deposit aquifers: Sand and gravel. Unconfined.	10 – 300	10 – 50	100	In most areas, deposits consist entirely of sand. Excessive iron concentrations.
Valley-fill deposit aquifers: Sand and gravel. Generally confined.	3 – 200	100 – 1,000	3,000	Glacial outwash and alluvium interbedded with clay and silt in many valleys are most productive water-bearing material in New York. Locally excessive iron or manganese concentrations.
Carbonate-rock aquifers: Limestone, dolomite, and marble. Unconfined in most areas.	10 – 300	50 – 150	200	Carbonate rocks are most productive bedrock unit in State. Water from this unit usually hard and contains hydrogen sulfide gas in some areas. From Niagara Falls to vicinity of Syracuse and in St. Lawrence valley, deep wells yield slightly salty water and, in places, water with a sulfate concentration that may exceed 300 mg/L.
Sandstone aquifers: Includes both sandstone and conglomerate. Confined in most areas.	3 – 500	50 – 100	100	Sandstone is the second most productive bedrock unit in New York. Water commonly slightly hard and has excessive iron concentration locally.

Long Island				
Upper glacial aquifer (includes Jameco and Port Washington aquifers): Outwash deposits (mostly between and south of terminal moraines but also interlayered with till) consist of quartzose sand, fine to very coarse, and gravel, pebble to boulder sized. Unconfined.	50 - 500	50 - 1,000	1,500	Main source of drinking water in central and eastern Suffolk County. Contains high concentration of nitrates and organic compounds in western Long Island. Saline water problems in extreme eastern end of Long Island.
Magothy aquifer: Sand, fine to medium, clayey in part; interbedded with lenses and layers of coarse sand and sandy and solid clay. Gravel is common in basal 50 to 200 ft.	150 - 1,100	50 - 1,200	2,000	Supplies most of the ground water for public-supplied drinking water in Queens, Nassau, and western Suffolk Counties. Saline water in North and South Forks and near Jamaica Bay.
Lloyd aquifer: Sand, fine to coarse, and gravel, commonly with clayey matrix; some lenses and layers of solid and silty clay; locally contains thin lignite layers and iron concretions.	150 - 1,100	50 - 1,000	1,200	Main source of drinking water for northwest shore of Long Island barrier islands to south. Saline water in North and South Forks and extreme west end of barrier islands.

SOURCE: U.S. Geological Survey, 1984.

TABLE 3.9 Primary Aquifers in Upstate New York

Aquifer	Community systems using the aquifer	Population using the aquifer
Schenectady	27	156,916
Endicott/Johnson City	30	110,457
Rampo/Mahwah River Valleys	3	74,500
Irondogenesee Buried Valleys	19	49,000
Jamestown	7	46,529
Big Flats/Horseheads/Elmira	8	40,350
Cortland	7	30,492
Corning	8	27,110
Olean/Salamanca	6	21,455
Fishkill/Sprout Creek	11	20,298
Clifton Park/Halfmoon	—	20,000
Owego/Waverly	8	16,535
Fulton	3	15,950
Seneca River	3	15,046
S. Fallsburgh/Woodbourne	2	12,000
Tonawanda Creek	3	12,360
Cohocton River	4	10,801
Croton-on-Hudson	1	8,100
		688,899

SOURCE: New York State Department of Health, 1981.

• "Principal" aquifers are underground formations known or suspected to be highly productive, but which are not heavily used for public water supply at this time. These are significant potential sources of future water supply.

Together, the primary and principal aquifers underlie roughly 10 percent of the upstate New York land area. The state's program includes policies designed to provide special protection for these areas. Recharge to upstate New York's ground water systems is derived from precipitation. Average annual rainfall ranges from 32 inches in the Central Lowlands and St. Lawrence Valley provinces to more than 50 inches in the Adirondack and Catskill regions.

Ground Water Quality Issues

While no major ground water quantity problems have occurred in upstate New York, there is evidence of ground water quality problems. Toxic contaminants, principally organic chemicals, provide the greatest cause for concern about upstate New York's ground water quality. Contamination from petroleum products and industrial and commercial solvents and degreasers has caused the closing of 24 wells serving 16 upstate public water supplies. Unlike Long Island, where 61 wells have been closed because of organic chemicals, upstate New York has not yet been widely affected by well closings. However, many of the closed wells are located in primary aquifers. The group of organics most commonly found are the halogenated organic solvents: trichloroethylene, tetrachloroethylene, and 1,1,1 trichloroethane. Numerous instances have been recorded of localized well contamination by gasoline and petroleum product constituents as well as other hazardous material leaks or spills. Petroleum product constituents are the most commonly reported type of organic contamination of private household wells. The bulk storage of materials is considered to be the state's most serious source of ground water contamination at the present time.

Nitrate contamination has been found in some upstate locations, but it does not appear to be widespread. Chloride contamination has been found in some private household wells (salt piles appear to be the primary cause). Recent findings indicate that pesticides could be a problem in upstate ground water in some locations, but the potential scope has not yet been sufficiently investigated. Other materials (toxic metals, other organic and inorganic substances, and radiological and bacteriological contaminants) do not appear to constitute severe or widespread ground water quality problems in upstate New York.

Ground Water Management and Protection

The two state agencies with responsibilities most directly related to ground water management are the New York State Department of Environmental Conservation (DEC) and the New York State Department of Health (DOH). DEC is the state's water resource manager and is directly responsible for development of the state's ground water program. Major elements of the DEC's water program that are integral to ground water management include water resources planning, ambient water quality standards and classification of ground water, water discharge permits, and programs that provide for the development, operation, and maintenance of waste water facilities. The DEC established a system of ground water classification and standards in 1967; the most recent revision was in 1978. DEC is now developing a petroleum bulk storage program to implement legislation enacted in 1983. The regulations for the program were adopted in late 1985.

Other elements of DEC's programs that provide for ground water protection include programs to regulate landfills, programs to regulate hazardous waste generation, storage, transport, and disposal, and the pesticide registration program. The state's *Draft Upstate Ground Water Management Program* report, released by DEC in early 1985, provides a comprehensive framework of recommendations for the management of these programs to achieve enhanced ground water protection.

Under the Public Health Law and Part 5 of the State Sanitary Code, DOH ensures that public water supply systems are operated properly and maintained to ensure a safe and adequate supply. The program involves regulation, periodic monitoring of water quality, inspection of systems, emergency response to problems of supply and quantity, laboratory services, and establishment of drinking water standards.

By agreement with the state, many county health agencies assist with state pollution control and water supply regulation programs. Some counties also administer additional programs of their own. This involvement of county government is a crucial element of effective programs to protect ground water. Towns, cities, and villages are responsible for regulating land use, a key factor determining the pollution hazard to ground waters.

LONG ISLAND, NEW YORK

Overview of Ground Water Resources

The largest and most important water resource in New York State is the vast aquifer system that underlies all of Long Island and includes Nassau and Suffolk counties as well as the boroughs of Brooklyn and Queens in New York City. It is the only source of fresh water available to more than 3 million residents.

Long Island is the only portion of New York lying within the Coastal Plain physiographic province. It is underlain by unconsolidated stratified glacial drift sand and gravel deposits atop a much thicker marine sands deposit, all of very high permeability. About half of the relatively high average rainfall rate of 44 inches per year percolates to the water table, providing an unusually high recharge rate to the system (under natural conditions).

The principal aquifers of Long Island (see Figure 3.1) are the upper glacial aquifers of Pleistocene Age and the deeper Magothy and Lloyd aquifers composed of older marine sands. The upper glacial aquifer in most areas is generally in direct contact with the underlying Magothy aquifer, but the deeper Lloyd sand aquifer is, with a few exceptions, separated from the overlying Magothy by a thick confining layer of silt and clay. The top two aquifers are pumped extensively for public and private supplies, but by far the most important of the three formations is the Magothy. Over 90 percent of all Nassau County withdrawals rely on this aquifer. Pumpage has in-

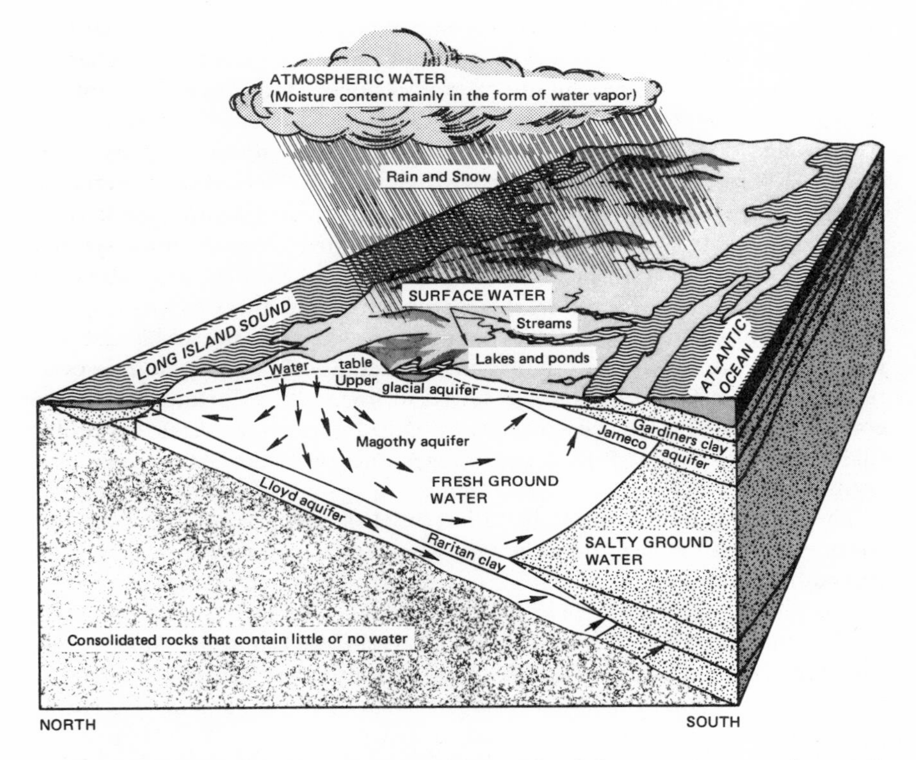

FIGURE 3.1 Generalized cross section of Long Island showing sources and types of water, major hydrogeologic units, and paths of ground water flow.
SOURCE: U.S. Geological Survey, 1982.

creased steadily since the 1930s. In Suffolk County, which covers the eastern two thirds of the island, most of the ground water withdrawal is for public supplies and private households. The agricultural areas of eastern Suffolk use ground water for irrigation. Similarly, most of the ground water use in Nassau and Queens counties is for public supplies. In Kings County, on the western end of the island, the principal use of ground water is for industry, and ground water use in general is rather limited.

Ground Water Quality Issues

Long Island's combination of relative low relief, high-permeability sediments, and high rainfall makes it especially vulnerable to ground water contamination. Those natural conditions combined with a large population, a significant amount of industrial activity, extensive use of agricultural and household chemicals including fertilizers and pesticides, widespread use of septic systems, leaking underground storage tanks, and heavy withdrawals of ground water have led to numerous and extensive ground water quality and quantity problems. From 1976 to 1981, some of the approximately 1000 major public drinking water wells were closed or restricted because of contamination from synthetic organic chemicals. In addition, nearly 1300 private wells were contaminated above the state guideline of 7 ppb by the agricultural pesticide, Temik.

In the upper glacial aquifer, salt water encroachment is a current problem on the islands and peninsulas of eastern Suffolk County and is a potential problem along all of Long Island's shores. Septic systems and agricultural and lawn fertilizers have resulted in increased chloride and nitrate concentration over extensive areas. Pesticides, industrial wastes, and landfill leachates have contributed to contamination problems of the uppermost aquifer in many areas.

The Magothy aquifer directly underlying the glacial aquifer also has salt water encroachment problems in some areas and current and potential contamination problems from organic chemicals, such as chlorinated solvents, in many locations. The deeper Lloyd aquifer is less subject to contamination by humanly introduced chemicals and also is affected in some areas by salt water intrusion. The Lloyd aquifer is by present policy generally restricted for use by only coastal communities that cannot use the shallower formations due to chemical or saline contamination.

Ground Water Quality Considerations

Nassau County, unlike Suffolk, has an added concern regarding the total volume of water withdrawn from the aquifer and consumed. Approximately half of all water withdrawn by public water suppliers is discharged after use into the sewer systems in the county. This water is treated and released into marine water, thereby lost for reuse through recharge. Additionally, ground water mining is occurring in those parts of the county with higher population densities. Over 50 percent of the county's 1.3 million residents rely on ground water mining practices. The consequences of excessive withdrawals include the loss of stream flows, the drying of lakes and ponds, a lowering of the water table, the increased spread of chemical contamination to deeper portions of the aquifer and the destabilization of the fresh water/salt water interface, leading to salt water intrusion.

Ground Water Management and Protection

Because of the high sensitivity of Long Island's ground water resources to all sources of contamination and the critical need for a continuous supply of high-quality ground water, many significant management and protective programs have been initiated over the past 20 years or so. These programs consist of several components involving local, regional, state, and federal agencies and organizations.

The Long Island aquifer system was one of the first to be designated a "sole source aquifer" in 1978 by EPA under provisions of the Safe Drinking Water Act. This designation, together with state regulations and policies, has provided the foundation for the development and implementation of a complex, comprehensive, integrated regional ground water management and protection program for Long Island.

The Long Island program includes elements covering the following general areas of program activity:

- Resource management—ambient water quality standards; aquifer classification system; monitoring program planning, review, and management.
- Source controls—hazardous material storage and handling; industrial/commercial/municipal waste management; sewage treatment; pesticide/fertilizer controls; underground storage tank restrictions.
- Zoning and land use controls.
- Water supply management—well permits, driller registration, public water supply.
- Drinking water—water quality guidelines for synthetic organic chemicals.
- Response and remedial actions—contamination response and Superfund programs; contaminated aquifer management; water treatment at well head; sewering to replace septic systems; regional water distribution and importation.
- Public education and participation.
- Regulatory enforcement.

Although more than 20 federal, state, regional, county, and local agencies and organizations are involved in carrying out this program, the leading regulatory roles are played by EPA at the federal level, the Department of Environmental Conservation and the Department of Health at the state level, and, at the county level, the Nassau and Suffolk County Health departments and the New York City Health Department.

A key element of the island's ground water protection program is the designation of extensive "hydrologic zones" that provide most of the deep flow recharge water to the deeper aquifers. These designated areas have been proposed to use highly restrictive regulations that would prohibit many types of industries and activities that could potentially release contaminants to ground water. The program includes special policies and regulatory requirements to provide especially stringent protection for the deep recharge zones. Full utilization of such controls is dependent upon the cooperation of the local towns, which have land use control powers but are reluctant to limit growth over such areas.

NEW JERSEY

Overview of Ground Water Resources

Ground water is very important in New Jersey and is used extensively throughout the state for public, industrial, domestic, and agricultural supply. Approximately 50 percent of the population of the state uses ground water as a source of drinking water. In 1980, about 730 million gallons per day of fresh water were pumped from aquifers in the state. However, areal and seasonal variations in ground water withdrawals can be significant.

The Coastal Plain is the largest physiographic province in New Jersey. The geology of the Coastal Plain is characterized by unconsolidated sand, gravel, silt, and clay thickening seaward from a feather edge at the Fall Line to more than 6500 feet thick in southern Cape May County.

The principal aquifers of the state are classified into two groups—Coastal Plain aquifers south of the Fall Line and non-Coastal Plain aquifers north of the Fall Line. The aquifers are described in Table 3.10 from the youngest to the oldest.

Ground Water Quality Issues

Major sources of contamination in New Jersey include septic tanks, municipal landfills, industrial landfills, surface impoundments, underground storage tanks, salt water intrusion, agricultural runoff, and pesticides. Other sources that have been identified are illegal dumping, leaky sanitary sewer lines, and abandoned wells. In October 1985, there were more than 400 active ground water pollution cases assigned to hydrogeologists in the Division of Water Resources, including 78 Superfund cases that involved ground water. Nearly 70 percent of the cases involved synthetic organic chemicals, primarily industrial solvents, and another 20 percent resulted from hydrocarbon discharges, including leaking gasoline tanks.

Ground Water Management and Protection

The New Jersey Department of Environmental Protection, Division of Water Resources, is the primary agency responsible for managing and regulating water resources in the state both for quality and for quantity. The New Jersey Water Supply Authority Act (1981), the Water Supply Bond Act (1981), and the New Jersey Water Supply Management Act (1981) are elements of the state program to protect and manage ground water sources. The New Jersey Water Supply Master Plan provides the framework for planning the water supply developments of the future. Although the state has regulated ground water withdrawals for public supplies since 1910 and other users since 1947, it was not until 1981 that all ground water users, including users of supplies exempt due to "grandfather rights," were required to obtain a permit and report monthly withdrawals to the Division of Water Resources.

Seriously depleted aquifers in the coastal areas are now subject to special management controls by their designation as Water Supply Critical Areas. Within such areas, in order to avoid excessive drawdowns or salt water intrusion, all existing wells are required to reduce their withdrawals by a specified percentage and to use surface water or exempted aquifers instead. Feasibility studies are conducted to develop these alternatives. Within such areas all withdrawals of over 10,000 gallons per day are controlled.

The Water Supply Bond Act provides a fund of $350 million for planning, designing, acquiring, and constructing water supply facilities as outlined in the Water Supply Master Plan and for ground water studies that do not involve construction. The New Jersey Geological Survey, an agency of the Division of Water Resources, and the USGS are currently engaged in aquifer studies covering the Atlantic City, Camden, N.W. Mercer, and South River areas and defining the extensive buried valley aquifer systems in the northern part of the state.

New Jersey has developed a state ground water discharge permit program pursuant to the New Jersey Clean Water Act of 1976. This program controls existing and future sources of contamination to ground water. About 2000 facilities will eventually require permits. Ground water quality standards and an aquifer classification system have been put into place to provide the basis for effluent limitations.

Under sponsorship of the Department of Environmental Protection, a long-range program for protection of aquifers is being developed by Rutgers University. Similarly, a demonstration project for aquifer protection is being carried out by Middlesex County, with the aid of a consultant retained by the state.

TABLE 3.10 Aquifer and Well Characteristics in New Jersey

Aquifer name and description	Aquifer withdrawals in 1980 (Mgal/d)	Well characteristics			Remarks
		Depth (ft)	Yield (gal/min)		
		Common range	Common range	May exceed	
Coastal Plain aquifers:					
Kirkwood–Cohansey aquifer system: Sand, quartz, fine to coarse grained, pebbly; local clay beds. Unconfined.	70	20 – 350	500 – 1,000	1,500	Ground water occurs generally under water-table conditions. Aquifer system extends from southern Monmouth County to Delaware Bay and from 12 mi southeast of the Delaware River to the Atlantic Ocean. Aquifer thickness can exceed 350 ft. Brackish and salty water may occur in coastal areas.
Atlantic City 800-foot sand: Sand, quartz, medium to coarse grained, gravel, fragmented shell material. Confined.	20	450 – 950	600 – 800	1,000	Principal confined artesian aquifer supplying water along the barrier beaches in Cape May, Atlantic, and Ocean Counties. Aquifer thickness generally ranges between 100 and 150 ft. Water quality suitable for most uses.
Wenonah–Mount Laurel aquifer: Sand, quartz, slightly glauconitic, very fine to coarse grained, layers of shells. Confined.	5	50 – 600	50 – 250	500	Important confined aquifer in the northeast and southwest part of the Coastal Plain. Aquifer thickness generally range between 60 and 120 ft. Water quality suitable for most purposes.
Englishtown aquifer: Sand, quartz, fine to medium grained, local clay beds. Confined.	12	50 – 1,000	300 – 500	1,000	Important source of water for Ocean and Monmouth Counties. Confined aquifer thickness generally ranges between 60 and 140 ft. Excellent water quality.

Potomac–Raritan–Magothy aquifer system: Alternating layers of sand, gravel, silt, and clay. Confined.	243	50 – 1,800	500 – 1,000	2,000	Highly productive and most used confined aquifer in the Coastal Plain. Aquifer system extends throughout Coastal Plain and attains maximum thickness of 4,100 ft. Includes two aquifers in northern Coastal Plain: Farrington and Old Bridge aquifers. Salty water increases with depth and in downdip direction. Excellent water quality but large iron concentrations in some areas.
Non-Coastal Plain aquifers:					
Glacial valley-fill aquifers: Sand, gravel, interbedded silt and clay. Generally unconfined except where overlain by lake silt and clay or till.	--	10 – 300	100 – 1,000	2,000	North of terminal moraine occur principally as channel fill in preglacial stream valleys; south of moraine, as outwash plains and valley trains. Important aquifers in Bergen, Essex and Morris Counties. Water quality suitable for most uses.
Aquifers in the Newark Group: Shale and sandstone: Shale, sandstone, some conglomerate. Unconfined to partially confined in upper 200 ft; confined at greater depth.	--	30 – 1,500	10 – 500	1,500	Most productive aquifers in Essex, Passaic and Union Counties. Water generally hard; may have large concentrations of iron and sulfate. Saltwater has intruded areas of large ground-water withdrawal near bays and estuaries.

SOURCE: U.S. Geological Survey, 1984.

Recently enacted legislation requires that industries selling property in the state must obtain approval from the department prior to sale. Approval is contingent on the cleanup of any hazardous waste including contaminated ground water. Since its inception, more than 1000 approvals have been sought; 200 sites have required ground water cleanups prior to sale.

Model programs to protect ground water have been developed at the local level, concentrating on recharge management in Middlesex County. Land management guidelines are being developed for incorporation into model ordinances.

WISCONSIN

Overview of Ground Water Resources

Wisconsin is a state that is heavily dependent on ground water and has been very active in developing programs to manage and protect ground water. About 70 percent of the state's population relies on ground water for its supply. About half of the total municipal supplies are from ground water, and all rural supplies are from ground water. Other uses of ground water include livestock use, which accounts for about 12 percent of all ground water withdrawal; industrial uses, which amount to approximately 11 percent of ground water withdrawals; and irrigation uses, which amount to about 14 percent. The largest withdrawals are for irrigation in central Wisconsin and for municipal supplies for Eau Claire, Janesville, La Crosse, and Madison. The natural chemical quality of the water throughout the state is suitable for human consumption and most other uses in most locations. The principal aquifers in Wisconsin consist of surficial glacial deposits and sedimentary sandstones and dolomites. Although the crystalline and igneous and metamorphic rocks that underlie these formations are used for local and individual domestic wells, the unconsolidated sand and gravel aquifer is one of the most important ground water supplies in the northern one third of the state. The Silurian dolomite aquifer is in a restricted area along the eastern coast of Wisconsin. Perhaps the most important aquifer in the state is the so-called sandstone aquifer, which underlies the southern two thirds of the state and is composed of sandstones and dolomites (see Table 3.11).

Ground Water Quality Issues

Wisconsin has experienced a number of problems in ground water contamination. The principal sources of concern include leaching of agricultural pesticides and fertilizers, underground storage tanks, municipal waste

TABLE 3.11 Aquifer and Well Characteristics in Wisconsin

Aquifer name and description	Well characteristics				Remarks
	Depth (ft)		Yield (gal/min)		
	Common range	May exceed	Common range	May exceed	
Principal aquifers:					
Sand and gravel aquifer: Unconsolidated sand and gravel; variable amounts of silt, clay, and organic materials. Thickness 0-600 ft; commonly 50-200 ft. Generally unconfined.	30 – 100	400	10 – 100	2,000	A well in Janesville was pumped at more than 5,000 gal/min. The water is very hard except in north-central Wisconsin. The median dissolved-solids concentration is 219 mg/L.
Silurian dolomite aquifer: Dolomite; some shale. Thickness 0-700 ft; thickest along Lake Michigan. Generally unconfined where shallow; confined where deep or overlain by clay sediments.	50 – 180	450	5 – 50	200	Important aquifer because it underlies the most densely populated part of Wisconsin. The water is commonly very hard. The median dissolved-solids concentration is 377 mg/L.
Sandstone aquifer: Sandstone, dolomitic sandstone, and dolomite; some siltstone. Thickness 0-2,700 ft thick in south; thickest in southwest. Confined in eastern Wisconsin by Maquoketa Shale; locally confined elsewhere.	50 – 1,000	2,000	10 – 500	1,000	Yields are commonly proportional to thickness of aquifer open to the well. The water is commonly very hard. The median dissolved-solids concentration is 307 mg/L.
Other aquifers:					
Precambrian igneous and metamorphic rocks; sandstone in northwest. Thickness unknown, but in thousands of feet. Generally unconfined where shallow; confined where deep or overlain by clay sediments.	50 – 100	400	0.5 – 10	50	Sandstone in northwest may yield 300 gal/min. Elsewhere yields generally do not exceed 50 gal/min.

SOURCE: U.S. Geological Survey, 1984.

One of the other major success stories for the Wisconsin ground water program has been a strong information and education program. Wisconsin has produced public information spots for television and radio, a ground water teacher's guide for elementary schools, posters on ground water and the hydrologic cycle, and several magazines on how ground water problems are dealt with by state agencies.

As mentioned above, the two-tiered standards approach by Wisconsin is the foundation of their protection program. The first standard, the "preventive action limit," serves two purposes. First, it must be used in the design codes for facilities such as landfills or other activities like pesticide applications so that contamination is prevented through the use of stringent design. Second, it serves as a trigger for remedial actions. If the preventive action limit is exceeded, some regulatory response may be necessary. In such a case the regulatory agency involved is required to make an evaluation of the particular problem and take appropriate action. When a preventive action limit is exceeded, a regulatory agency may prohibit continuation of the activity causing the problem. The regulatory agency must assure restoration of ground water quality to below the preventive action limit if it is technically and economically feasible.

The second standard is the enforcement standard. When substances are detected entering ground water at concentrations equal to or above its enforcement standard, a violation is said to have occurred and the activity, practice, or facility that is the source of the contamination is subject to immediate enforcement action. Unlike the preventive action limit, where technical and economic feasibility are considered, when an enforcement standard has been attained or exceeded, a regulatory agency must prohibit the continuation of the activity from which the substance came unless it is demonstrated to the agency that an alternative response will achieve compliance with the enforcement standards.

4
State and Local Strategies to Protect Ground Water

This chapter contains the bulk of the information that the committee has collected and evaluated on each of the different strategies and components of the state and local ground water protection programs described in Chapter 3. This is not a comprehensive inventory and assessment of all ground water protection programs currently in place in these states and local areas. Each of the state and local programs was examined for particular characteristics, which, when viewed collectively, cover most of the major approaches to ground water protection that have been attempted across the United States.

The committee classified ground water protection program approaches into five major categories:

1. Information Collection and Management Systems
2. Classification Systems
3. Ground Water Quality Standards
4. Control of Contamination Sources
5. Implementation of Ground Water Protection Programs

INFORMATION COLLECTION AND MANAGEMENT SYSTEMS

A successful program must be founded on an information base that allows proper definition of problems and evaluation of prevention strategies. A prevention program should be based on adequate surveys of (1) water resources and their location, (2) ground water basin characteristics with respect to the potential for contamination, and (3) present and anticipated land and surface water uses that can affect ground water. Information on water resource conditions and new ground water research must be easily available to decision makers. Table 4.1 summarizes the components of such an information base. Some of the needed data base components are available for most states, perhaps obtained as part of a ground water management program, but more often the data have been gathered for other purposes. Some states have a good understanding of the hydrogeology of their ground water basins, but for others such information is fragmentary. Some ground water systems are much more complex than others. The degree to which a sound management program can be structured will rely on the depth and breadth of the information base available. In this chapter, examples are given of the development and use of information bases to formulate ground water quality management policies and control strategies.

While all the information components listed in Table 4.1 are useful in an overall ground water quality management program, each aspect of a program may draw more heavily on one or only a few of the components. In this section, each of the four main components in Table 4.1 is described briefly to serve as examples of how they play a significant role in the development of an overall ground water management program. Each of these properties is complex and difficult to quantify.

Hydrogeology

One of the most basic needs of a ground water quality management program is an understanding of the system's hydrogeology. This will help determine the water yield characteristics of the aquifer; the suitability of the water, in terms of quality, for different beneficial uses; the degree of vulnerability of the aquifer to contamination at different locations; and the necessity for a stringent program of contamination control.

The characteristics of soils overlying an aquifer play a significant role in the potential for aquifer contamination. Aquifers overlain by permeable sands or gravels are highly vulnerable to surface contamination since contaminants can move rapidly through such materials. Clay, on the other hand, is rather impermeable and can retard contaminant movement, providing more time for corrective action on chemical spills. Knowledge of the boundaries of ground water basins and the characteristics of the aquifers themselves provides important information on the limits to which a contaminant may spread and the particular areas that are vulnerable to contamination from individual sources.

With respect to aquifer characteristics, shallow unconfined aquifers are highly vulnerable to contamination compared with deep aquifers that are

TABLE 4.1 Information Base Components for Ground Water Management Decisions

Hydrogeology

- Soil and unsaturated zone characteristics
- Aquifer characteristics
  - Depths involved
  - Flow patterns
  - Recharge characteristics
  - Transmissive and storage properties
  - Ambient water quality
  - Interaction with surface water
  - Boundary conditions
  - Mineralogy, including organic content

Water Extraction and Use Patterns

- Locations
- Amounts
- Purpose (domestic, industrial, agricultural)
- Trends

Potential Contamination Sources and Characteristics

- Point sources
  - Industrial and mining waste discharges
  - Commercial waste discharges
  - Hazardous material and waste storage
  - Domestic waste discharges
- Nonpoint sources
  - Agricultural
  - Septic tanks
  - Land applications of waste
  - Urban runoff
  - Transportation spills (may also be considered a point source)
  - Pipelines (energy and waste water) (may also be considered a point source)

Population Patterns

- Demographic
- Economic trends
- Land use patterns

protected by overlying impermeable layers. Flow patterns indicate directions that contaminants may move within an aquifer. Potential yields of water from an aquifer indicate the size of the water supply that may result from full aquifer development, giving a measure of its potential value to the region.

At the state level, hydrogeologic data are used most frequently in permitting and siting decisions for individual facilities. Only in a few states has hydrogeology provided the basis for structuring a comprehensive management program. Connecticut has mapped its stratified drift deposits where there is good potential for water supply development and structured its classification system based on ground water basin mapping. Vermont and New York State have mapped valley fill aquifers in connection with their state planning programs, primarily for the purpose of defining potential water supply aquifers. Florida utilizes hydrogeologic data on a regional scale by distinguishing between confined and unconfined aquifers in its classification system, although the aquifers are so large and vertically arranged that the system provides more guidance on injection well disposal than land use or siting.

California has mapped its ground water basins in connection with its basin planning program, for the purpose of establishing beneficial uses and ambient water quality standards. Although differential classifications based on beneficial use are possible, as a practical matter, virtually all the basins are designated municipal drinking water supplies.

Local programs have made far more extensive use of hydrogeologic data than have state programs. Long Island, New York, is perhaps the best example of the effective integration of hydrologic information into a ground water management program. Regional flow and interaquifer connection form the basis for the hydrogeologic zoning plan developed for Nassau and Suffolk counties, Long Island, through a Section 208 planning grant under the Clean Water Act. This information has been used to determine those aquifer recharge areas where a potential source of pollution must be most carefully controlled. A host of local land use controls, county-level ordinances, state laws, and regulatory programs is coordinated on the basis of this single integrated approach.

Local programs in several locations have used hydrogeologic data to define zones of influence for public supply wells as the basis for land use zoning. The Cape Cod Planning and Economic Development Commission has calculated areal zones of contribution for existing and planned future public supply wells. Local communities have used this information for development of zoning overlay protection by-laws. In Florida's Dade and Broward counties, data on zones of influence provide information on public supply wellfields in the Biscayne aquifer. County and local ordinances have been adopted based on travel time to the wells within the zone of influence. Principally in the northwestern and midwestern states, local communities have developed aquifer protection plans based on hydrogeologic mapping with identification of aquifers, recharge areas, and wellfields.

The Long Island hydrogeologic zoning system is based on an extensive data base accumulated through decades of study of a complex flow system. However, the remainder of the above examples have been accomplished in areas of relatively simple aquifer systems without extensive data gathering. In many areas the available data should be adequate to develop at least first-draft maps suitable for the enactment of zoning and protective ordinances. As data are gathered over time, the areas can be revised. In fact, all the programs previously discussed are subject to periodic revisions, including realignment of the boundaries delineating critical protection areas. Although any change in a planning tool can generate controversy, the ongoing nature of programs to define aquifer hydrogeologic characteristics to date has not served as an impediment to passage or enforcement of effective controls.

The Massachusetts Department of Environmental Quality Engineering (DEQE) has developed a unique water supply protection atlas for use by state and local governments in ground water protection programs. The atlas consists of four overlays for each of the 177 USGS topographic quadrangle maps (scale: 1:25,000) that cover the state. The overlays consist of (1) sources of public water supply; (2) waste sources including surface impoundments, hazardous waste sites, landfills, auto junk yards, road salt storage areas, and permitted discharges to surface and ground water (1600 statewide); (3) aquifer information including areas of equal potential well yield from USGS atlas series; and (4) drainage basins delineating major river basin and subbasin divides. Each Massachusetts community has been issued the overlays to be used on the USGS quadrangle maps specific to their town boundaries. A handbook describing the atlas and providing instructions on verifying, updating, and expanding the information in the atlas has been distributed. A few towns have added contamination sources, such as underground storage tanks. DEQE is in the process of computerizing the system in conjunction with USGS. A sample portion of a quadrangle map with overlays is shown in Figure 4.1. The water supply protection atlas has assisted municipal government in land use planning and water supply management. DEQE has used the atlas in identifying priorities for monitoring enforcement and remedial action programs for its solid and hazardous waste regulatory program.

FIGURE 4.1 U.S. Geological Survey quadrangle map section of Massachusetts showing river basins, aquifer boundaries, public sources of water supply, and waste sources. SOURCE: Commonwealth of Massachusetts, 1982.

discharges, sanitary landfills, and inadequate facilities for toxic waste disposal. Although metallic mining is not a significant activity in Wisconsin, much of the state's concern and action in ground water regulatory actions was prompted by interest in metallic mine deposits and a potential for mining in the state. The second major incentive for a stronger ground water protection program was stimulated by the discovery of aldicarb and other pesticides in ground water and by controversies over landfill activities and operations. Wisconsin is a state with a long history of environmental concern and public awareness of the importance of environmental quality, specifically the importance of ground water quality. Therefore, there has been strong public involvement and interest in developing a strong and effective ground water protection program.

Ground Water Management and Protection

Wisconsin's ground water quality protection program is relatively new. It is based primarily on legislation passed in 1984, which included five main components. The first and foremost of these components is the development of ground water quality standards. The second is to provide funds for replacement of contaminated water supplies. The third aspect is to provide an environmental repair fund, and the fourth component is to develop a water quality monitoring network. The fifth aspect of this program is to certify laboratories to be used to analyze ground water quality. The Wisconsin program does not rely on a system of aquifer classification.

One of the main strengths of the program seems to be the traditional regulatory approach to managing water quality based on water quality standards. This is done in a more comprehensive approach than in most other states. Several things are unique about Wisconsin's ground water quality standards: first, they apply to any sort of environmental activity that is regulated by a state agency; and second, they have a two-tiered approach—for each regulated compound there are two concentration limits, one called the "enforcement standard," similar to a health advisory limit or a maximum contaminant level, and the other called a "preventive action limit," which is a fraction (10, 20, or 50 percent) of the enforcement standard. Another unusual aspect of Wisconsin's system is a compensation program for people with wells that become polluted with man-made chemicals. The state has a no-fault program that will pay for 80 percent of the cost of replacement of their well regardless of the cause of the pollution.

The Wisconsin program involves several state agencies, and an important part of that program includes the Ground Water Coordinating Council. This council consists of representatives from all the involved agencies who meet routinely to coordinate state ground water activities.

Ambient Water Quality

The ambient water quality, while listed as only one item under aquifer characteristics in Table 4.1, deserves particular emphasis because of its special significance in formulating a strategy for ground water management. For example, the state of Colorado has based a classification system for aquifers partially on the mineral quality because of its relationship to its suitability for various beneficial uses. The classification system formulated by the state of Connecticut is, to some degree, also related to the quality of the water in the aquifer.

In addition, knowledge about the degree of chemical contamination in aquifers can help to indicate areas of high vulnerability to aid in decisions about the type of regulations needed for contamination control as well as cleanup. Water quality data resulting from the long history of contamination on Long Island, including contamination from septic tanks and agricultural and industrial practices, has led to regulations or guidelines on sewering, density of residential land use, septic tank cleaner use practices, and underground storage of petroleum products and chemicals. Similarly, knowledge of the vulnerability of ground water to contamination from leaking chemical and petroleum product storage tanks, obtained through surveys of ground water quality, has led to the development of storage tank regulations in Santa Clara County, California, and in Cape Cod, Massachusetts.

Water quality data bases have traditionally contained information on concentrations of a limited number of inorganic ions. For the purpose of contamination control, ground water with potential for contamination should be analyzed for a range of organic contaminants that have frequently been found, such as pesticides, organic solvents, and petroleum components. The lack of a good water quality data base has made decisions about ground water contamination control difficult for most states. Because contamination is frequently local and may move quite slowly in an aquifer, sampling for chemical contamination needs to be extensive and carefully done.

In order to obtain adequate information on ambient water quality, the state of Florida is now developing a comprehensive ground water monitoring network to provide information about the quality of its polluted and pristine ground water. This testing program will include monitoring for the EPA list of 129 priority pollutants and provide baseline information that will be useful in evaluating ground water protection and restoration needs as well as in evaluating the effectiveness of prevention and restoration programs.

The state of California has a similar comprehensive monitoring program underway, pursuant to Assembly Bill 1803 of 1983. The first phase of the program, which is to be completed in 1986, will cover "large" water systems, defined as containing more than 200 connections. So far, the data show that about 12 percent of the more than 2500 sampled wells have detectable concentrations of organic chemicals found on the priority pollutant list, plus a supplemental list prepared by the state of California. About 4.5 percent of the wells had pollutant concentrations exceeding a state "action level" or federal maximum contaminant level. Generally, the most frequently found chemicals were the volatile organics, typically industrial solvents. The large volume of data generated will help state and local governments make land use decisions, formulate control regulations, and develop aquifer classifications.

The California monitoring program has also identified some difficulties to be overcome, as the program is extended to cover small water systems. First, quality control in participating laboratories has required a great effort. Compliance with analytical protocols specified by the California Department of Health Services (DOHS) was difficult to obtain, and a number of analyses had to be repeated in the interest of quality control. Second, data management, and especially quality control of entry and retrieval, has been identified as a topic of special concern to the program.

While many states use total dissolved solids (TDS) concentration as the principal criteria to define drinking water aquifers, few have incorporated actual water quality data into state ground water management plans. As with hydrogeologic information, local programs have made more use of water quality data to delineate ground water management zones.

New Jersey has used actual ambient water quality in a regulatory program. An extensive data gathering effort was conducted to characterize the ambient quality that supports the unique ecosystem (the Pine Barrens) in this area. The resultant values for pH, nitrate, nitrogen, and phosphorus were incorporated into the state classification system.

Presumed water quality is used to define class GB waters (not suitable for potable use without treatment) in the state of Connecticut. In this case, land use patterns and the presence of known discharges in an area were used as evidence of probable degradation of the ground water to a point where its use for drinking water supply would require treatment.

Both Nassau and Suffolk counties, Long Island, have their own monitoring networks, as well as assistance from the USGS in that area. Through this extensive water quality data base, ground water quality has been mapped in three dimensions based on nitrate and organic pollutant concentrations. This information was used to subdivide the hydrogeologic flow regions into areas of high-quality water and areas of existing contamination. Additional data on land use, including high-density residential development and industrial use, were employed to identify areas where contamination is likely.

County monitoring programs on Cape Cod and in Suffolk County, Long Island, use data from private well testing programs to provide regional information on ambient quality in the shallow aquifer. Each county maintains a computerized data base with more than 5000 private analyses entered as of 1985. The data are useful in correlating land use with water quality as well as identifying new problems.

Both the Cape Cod and the Suffolk County private well data bases have already identified problem areas where unreported chemical and oil spills have occurred or septic tanks have caused excessively elevated nitrate levels. The data have been used to make decisions for establishing priorities for water main extensions and revising land use controls. Investigations triggered by private well analyses have led to the discovery of leaking underground storage tanks in several instances.

Because the Cape Cod and Suffolk County private well data bases are maintained in the county agencies involved in ongoing water quality planning programs, they are effectively utilized in the planning process by these two counties.

Water Extraction and Use Patterns

A key component in making ground water quality decisions is knowledge of the state's ground water withdrawals and their use. This is critical in parts of the country where the ground water is being depleted at a rapid rate. The quality of water is likely to change with the lowering of the water table, particularly in the arid West. Cones of depression will alter the flow of contaminants. Shallow wells are more likely to be influenced by pollution. Therefore, to get a comprehensive picture for management of ground water quality, a state needs to know the location of water wells, the amount of water being extracted, and for what purpose it is to be used.

Arizona, for example, has a good data base for ground water withdrawal quantities. Considerable work has been done by the USGS in providing data over the years. In the developed areas of the state, depth to ground water is well documented. In 1980, Arizona passed a comprehensive Ground Water Management Act for the purpose of reducing the overdraft of ground water. Although the act initially addressed water quantity in its first management plan, ground water quality will be a major factor in the second (1990) plan, and management of both quantity and quality will be integrated.

Since 1982, withdrawal wells in Arizona have had to be registered regardless of their use. All wells within designated management areas, with the exception of wells used for domestic or stock watering purposes, have to be metered and the amount withdrawn reported yearly. By 1990, each category of water user—municipal, industrial, agricultural, mining, for example—must limit its water usage to the amount set for each section by the Department of Water Resources. As a result, well production information will be readily available.

Besides knowledge of extraction that is occurring, states need to be aware of activities relating to water reuse and artificial aquifer recharge. Increasingly, in the parts of the country where water supplies are being depleted or contaminated, water reuse and recharge are becoming economically viable and attractive. In the Phoenix and Tucson, Arizona, Active Management Areas, for example, virtually total reuse of water supplies is planned to occur by 2025. Any state where effluent is being reused and artificial recharge is taking place should be alert to the impact on ground water quality and provide necessary protection.

In Florida, ground water allocations are managed by the seven regional water management districts. The districts administer a permitting system and collect extensive water use data. At least one district, the South Florida Water Management District, has a computerized data base, with electronically monitored water levels in the vicinity of wellfields. Computer-generated maps showing zones of influence around pumping wells can be generated nearly instantaneously. This system was recently used to develop maps of zones of influence for submission to the state in a petition for sole-source aquifer designation under the new classification system.

Because of the advanced nature of the data base, the state of Florida is considering modeling zones of influence for all public wells as a means of implementing its classification system. These areas, as well as critical recharge areas to be mapped by the state, would be classified in the highest category for use and protection. If implemented, this system would be the first state-mapped wellfield protection requirement implemented through a classification system. It would also be the first state-level system utilizing hydrogeologic flow system data in this context.

Potential Contamination Sources and Characteristics

In order to protect ground water quality, it is important to know the pattern of production or use of potential contaminating substances. To be most useful, a data base on potential sources of pollution should contain information on (1) the quantities and chemical composition of the material, (2) the type and spatial location of its use, (3) the industry or reason for its use, and (4) the time of use. This information can be compared with data on spatial distribution of sensitive hydrogeologic zones and on water supply systems to help target efforts on source reduction, aquifer protection, water quality monitoring, and detection of improper waste disposal.

Maintaining a data base on the distribution of chemical use and disposal is essential for determining the type and location of sampling for chemical contaminants. The high cost of chemical analysis prevents routine monitoring for all potential contaminants. If a contamination event is detected, a data base on distributed use and disposal enables an efficient and speedy allocation of resources to determine the location of other possible contamination events. A use and disposal data base combined with chemical mobility information can be valuable in establishing chemical monitoring priorities for each location.

It is clear that the passage of the required laws and regulations and the ability to implement those laws are necessary to develop a good data base on chemical use and disposal. Potential ground water polluters, both private and public, typically resist the imposition of additional environmental and health regulations including data on chemical use. However, the attitudes of potential polluters have tended to change as it is recognized that the costs of cleanup of contaminated ground water can be very high. In the long run, collection of an adequate data base helps to protect sources of ground water discharge from the imposition of potentially large economic burdens by basing control programs on sound information.

An example of an extensive data base on chemical use is California's system on pesticides. This program applies only to those chemicals that are designated as restricted use pesticides. This program has been in existence since the early 1970s and currently includes about 80 pesticides. As discussed later in this chapter in the section on nonpoint source contaminants, the pesticide use data base has enabled California to determine which pesticides need to be monitored for in each location. Because of the large number of pesticides in use and the high cost for analyzing each one, the use data allowed the development of a cost-effective monitoring program.

Spatially distributed use data are also useful for other issues besides pesticides. For example, estimates of fertilizer use in Long Island indicated that residential use of fertilizer rather than septic tanks, as had been originally thought, is the major cause of nitrogen contamination of water supply wells.

Information on the industry (or in the case of agriculture, the crop) that is responsible for the use of the material is also helpful in directing efforts at source reduction. The alternatives for treating waste before disposal or for replacing the material by a less hazardous chemical depend on the production process and purpose for which the chemical is being used.

Properties of Chemicals

Another important aspect of a data management program is to obtain a listing of the properties of chemicals. The properties of these chemicals will be useful in determining the extent of ground water contamination, methods for aquifer restoration, and implementing regulations concerning chemical usage and control. The major properties of interest are listed in Table 4.2.

The rate of movement and fate of chemicals in the subsurface environment are affected by its physical, chemical, and biological characteristics. For example, trichloroethylene (TCE) moves more readily in an aquifer than does DDT. This is because of its lower sorption coefficient, which can be related to the chemical's higher water solubility and lower octanol/water partition coefficient. TCE is relatively nonbiodegradable under aerobic

TABLE 4.2 Properties of Chemicals of Interest to Ground Water Management Programs

Physical Properties

- Water solubility
- Octanol/water partition coefficients
- Vapor pressure (Henry's law constant)
- Sorptive characteristics
- Density
- Viscosity

Chemical Properties

- Structure
- Isomeric forms, homologs
- Transformation potential and end products
- Other commingled compounds and carriers

Biological Properties

- Biodegradability
  - Aerobic
  - Anaerobic
- Metabolic products

Toxicologic Properties
(The Chemical and Its Transformation Products)

- Acute effects
- Chronic effects

ground water conditions but can be transformed into dichloroethylene, vinyl chloride (a human carcinogen), and other chlorinated compounds. It can also be mineralized to inorganic products under anaerobic conditions. Such knowledge is essential in attempts to estimate the environmental significance, determine the original source of the contamination, and formulate cleanup strategies. For materials that sorb to soils, the mass of contaminant contained in aquifer water may be small in comparison with that sorbed to soil. Removal of the sorbed contaminant as well as that dissolved in the water must be undertaken in order to restore the quality of aquifer water. The best removal and ground water treatment procedures are also related to the properties of the contaminants.

Population Patterns

The impacts of projected population increases or decreases and industrial and land use changes are an important element in the management of water quality. There is a potential for dynamic change, e.g., snowbelt-sunbelt shift in population and industrial relocation and growth of high-tech industries versus heavy manufacturing.

Planners use this information as they look at the carrying capacity of the aquifers and the availability of proper waste disposal for both municipal and industrial waste. Certain types of uses can be matched to existing water quality (if there are no legal restraints), thereby reducing treatment costs. For example, irrigated farmlands can be expected to have contributed a variety of organic and/or inorganic chemicals to the vadose zone or to the water table. If this land is converted to urban use, treatment of the ground water for human consumption can be expected.

With population increases, the demand on an aquifer can result in the aforementioned changes in water quality from pumping alone. (See previous section on "Water Extraction and Use Patterns.") Large new home developments using septic systems can cause considerable water contamination problems. An influx of newly developed industries can be responsible for contamination of ground water through inadvertent spills and leaking petroleum and chemical storage tanks. On the other hand, where there is a decrease in population growth or shift in industrial location, some previous contamination problems may be reduced.

Data Processing and Analysis

Data Collection and Processing

Because of the diffuse nature of ground water sources and of the location of potential polluting activities, it is important to have a data base that describes the distribution of pollution sources and of the ground water resource. Given the large number of chemicals and the need to have an accurate spatial distribution, management of this kind of data base requires computers for data storage, retrieval, and analysis. With this system, data from many sources can be integrated and presented in forms that are concise and understandable.

Setting up a computer-based data management and processing system is complicated and expensive. However, the state and local ground water protection program representatives interviewed by the committee felt that the long-term benefits outweighed the costs because they would have a much better basis for allocating scarce resources and for making regulatory decisions. In many states a number of different agencies are collecting data that are relevant to ground water contamination. It would be useful to have a central repository for these data, so that the pieces of information processed from different sources are in a compatible form to facilitate comparison. The rapid decline in cost for powerful computing systems and the improvements in software encourage the wider use of computer-based data processing systems for ground water protection. No state reviewed by the committee has a comprehensive computerized program that integrates source, hydrogeologic, water supply, and monitoring data. However, the committee expects the more progressive programs to be moving in this direction. Following are specific examples of data bases developed in several states reviewed by the committee.

California This state is developing a system that will develop county maps describing the pattern of (1) restricted pesticides use, (2) pesticides detected in ground water, (3) soil types, and (4) depth to ground water. It is expected that the overlay of this graphical information will help both local and state officials to identify potential contamination events so that action is taken before the problem becomes overwhelming. This program has just begun and currently produces only maps that indicate whether any restricted pesticide has been used and whether any pesticide has been detected. This type of information can be overlain with hydrogeologic and water supply information to estimate the potential range of contamination that may be associated with a given activity.

There are, of course, problems with large data bases whether or not they are computerized, but some of these problems can be alleviated by modification of a computerized system. For example, the pesticide use forms in California are filled out by clerks in the office of the local agricultural commissioner. They write in numbers on the forms that give the coordinates of the location of pesticide use. There are some errors in this information as evidenced by the reporting of some pesticide use in coordinates that correspond to areas in the Pacific Ocean. However, the pesticide program expects to install microcomputers in the local agriculture commission's offices so that clerks will key the information directly into the computer. Besides saving on the expense of having manual data key punched later, this proposed system should reduce the amount of errors in data entry by programming the computer to give an error message if a clerk enters coordinates that are not in California.

The California pesticide program is also expected eventually to collect this information to assist in registration of pesticides. The state and the local agricultural commissioner determine which pesticides may be used in which areas. For example, aldicarb is allowed to be used in all except one county in California. They hope to screen out pesticides that may pose a problem in certain areas. The models require information on the chemical characteristics of the pesticide (water solubility, solids absorption, and half-life, for example) as well as information on the soil type, precipitation pattern, and irrigation.

Suffolk County, Long Island Suffolk County has developed an innovative and simple system of coding data for entry into a computer. Federal funds from the Department of Housing and Urban Development (HUD) were used to code existing data according to census geography GEO codes. Data can be retrieved by street address, census area, town, or other political subdivision. These data can be used by simple reference to code maps on file at the county. This eliminates the extra step of converting a location to latitude/longitude coordinates, which is required to use the USGS WATSTOR data base. Suffolk County's system is simpler and more accurate and provides more flexibility in data retrieval. The data base is maintained on a System 2000 Time Sharing Network maintained by the State University of New York at Stony Brook.

Caution must be used in attempting to estimate the magnitude of the temporal trend of a contamination problem by analyzing data that were collected ad hoc. For example, private well data can provide valuable information but may display a statistical bias. Residents tend to have their water analyzed if they suspect they have a problem or if they have heard that their neighbor has a problem. This is offset to a minor degree by the requirement that every new well must be analyzed prior to occupancy. An additional problem is that each analysis is likely to generate a single data point. Few wells are resampled. Thus, this kind of data base must be used with caution to monitor changes in water quality over time. The maintenance of fixed monitoring points for periodic analysis in conjunction with such a local program is necessary for true temporal analysis of water quality changes.

Kansas A major shortcoming in Kansas water data is the lack of accurate and current information on water use. Data for municipal and industrial use are reasonably current and accurate, but irrigation pumping data are not, because of widespread resistance to metering.

Kansas follows a grid system of mapping that designates the exact location of each gas, oil, water well, industrial activity, or storage site. Its ground water basins have been mapped, and the chemical quality (inorganics) is known, along with appropriate boundaries and depths. Production data are available for municipal and industrial wells and for some irrigation wells. Most Kansas counties have ground water and geology reports, some of which date from the mid-1930s. For more than a decade, the reports have been based on aquifer systems, of which the Ogalalla is the most pervasive and intensively studied.

Considerable information is available on amounts and areas of aquifer recharge, and the state has studies embracing recharge areas. Scans of organic compounds have been done at a small percentage of municipal wells, and an expanded program of screening for purgeable organics is under way. Laboratory capability for organics, while substantial, is being expanded, and private analytical laboratories are licensed by the state.

Data Measurement and Analysis

Water quality and hydrogeologic data needs can be met only by adequate laboratory facilities. These facilities must produce data for prevention as well as for the urgent demands generated by emergencies and pollution cleanup.

Laboratory capability may be available from local health departments and large water utilities, state health and natural resource agencies, USGS, EPA, and private facilities.

Local and state laboratories are usually depended on for meeting the continuing needs for data for ground water protection and remediation programs. The demands on them are urgent and compelling when data are needed for chemical and oil spills and hazardous waste cleanup, often resulting in a lower priority for analysis of ongoing and prevention-related water and soil samples. Many of these laboratories have a continuous program of education and quality control to assure accurate results. Nearly all states with large analytical loads can benefit from in-house laboratory capability to check outside analyses and to provide other special analytical services on demand.

Adequate funding for laboratory equipment and personnel and the capacity to manipulate and analyze results are essential for an effective state ground water protection program. Funding sources need to be stable and as continuous as possible. Dedicated fees, contracts with large water utilities, general revenue, federal program funds, and local taxes, or some combination thereof, are used to obtain laboratory funds. Some states have used matching grants and contracts with the USGS. Most of the states call on the regional EPA laboratories for assistance from time to time with special problems. However, assistance from federal laboratories does not eliminate the need for a state laboratory.

Many states do not have the capacity to monitor organics in ground water without assistance from private laboratories. For that reason states such as Kansas, California, and Wisconsin have turned to the certification of private laboratories. While numerous examples could be given of state certification programs, programs from Kansas and Wisconsin are discussed here.

Kansas In 1974–1975, Kansas began to certify laboratories for specific procedures required for the analysis of water and air. Two professionals manage the program with the equivalent of half-time travel for each to make on-site evaluations. In addition to private laboratories, those of large public utilities are certified for some tasks.

The Kansas laboratory is supported by fees and general state revenue. In addition, the state has used a substantial part of its EPA Safe Drinking Water Act funds to buy equipment. The capacity for analyzing volatile organic chemicals has increased 30-fold in two years. Laboratory expansion was included in the State Water Plan as well as in the program documents of the Department of Health and Environment. Being included in the plan has made funding for new equipment easier to obtain.

Much information is available on inorganics in ground water, but knowing where to find it may be a problem. The state geologist has tried unsuccessfully to get legislative support for developing a comprehensive, integrated data collection program. Kansas is moving toward a decentralized data system with a central contact point and toward a system to give feedback on program effectiveness. In general, most data for recent years are available from WATSTOR or STORET, the two national data bases for chemical quality.

Kansas counties have been required to have solid waste disposal plans for 15 years. Most use landfills for disposal, and those judged most likely to leak have monitoring wells. The oil and gas data base is much improved through industry-funded contracts and support. A contractor supplies data and analysis for all oil and gas drilling in the state and compares them with similar data from the other oil-producing states. Historical data begin in the 1950s, with some variation in startup dates between the counties. The records provide extensive data on stripper wells (those that produce less than 10 barrels per day), projecting trends, giving types of wells, showing changes in production and sales, and analyzing taxes paid. Stripper wells produce the most salt water for each barrel of crude, yet produce the least income. Hence, they require the most surveillance to assure that brine does not get into the fresh water. This information combined with the well completion and field data not only benefits the industry but aids programs to protect the fresh water.

Kansas does not have data to show the added load that requirements for clean air, clean streams, and solid waste programs have placed on the soil and in some cases the ground water. In Kansas, fumigants (carbon tetrachloride) at grain elevators and cleaning solutions (trichloroethylene or tetrachloroethylene) are the most frequently found organics. Limits have been set for 24 volatile organics.

Wisconsin Wisconsin has recently proposed administrative rules establishing guidelines for laboratory certification and registration. In their two-tiered water quality standards system, laboratory certification is the more stringent requirement. Laboratory registration applies to "in-house" laboratories only, i.e., laboratories not providing testing services for hire. The proposed administrative rules require certification and registration of laboratories where samples are collected and analyzed for the following:

- Solid or hazardous waste facilities
- Mining permit applications
- Waste water permits
- Compliance with ground water quality standards
- The Safe Drinking Water Act program

Commercial laboratories performing tests associated with these activities will have to be certified and follow minimum criteria for test methodology, quality control, and quality assurance. Registration of laboratories performing tests on an in-house basis may be filed with the Department of Natural Resources (DNR) if compliance is demonstrated with criteria similar to those required for certification. Certification and registration in Wisconsin will be renewed annually, and guidelines will exist for revocation or suspension of certification or registration.

In order to oversee this program, a committee of persons outside the DNR is established to review certification and registration procedures and to make recommendations to DNR regarding test methodologies and standards for certification and registration. This committee consists of appointed representatives of municipalities, industrial and commercial laboratories, public water utilities, and waste disposal facilities. In addition, rules relating to laboratory certification developed by either of the two state regulatory agencies responsible for developing water quality standards must be approved and recognized by both agencies.

Conclusions and Recommendations

Hydrogeologic Information

Protection of ground water requires a sound and appropriately designed hydrogeologic information base to determine on a continuing basis what ground water contamination problems may exist. Data are also needed to predict future threats.

The committee encourages state and local programs to obtain the necessary hydrogeological information for each region. The program should be long term to obtain physical and chemical information aimed at developing a quantitative understanding of the occurrence and the quality and dynamics of the resource, together with the types, extent, and sources of potential contaminants. The data should be collected and formatted to assist in the area's ground water management program so that the program's effectiveness over time can be evaluated. The USGS should expand its technical assistance and information-gathering programs to assist states in this effort. State and local organizations should become familiar with and incorporate appropriate data available from federal systems, such as those of the USGS and the Department of Agriculture, relating to hydrology, soils, and chemical use.

Types of Data and Data Management Systems

Recent advances in electronic data storage and processing technology have enabled collection and management of large amounts of data. This has often encouraged data gathering without adequate assessment of its usefulness and without conversion of the data into readily usable formats for analyses, policy-making, and management.

The committee recommends that both state and federal information programs be carefully designed to emphasize collection and storage of data that can be produced in a format that facilitates analysis of problems and long-term trends. The programs should be reviewed and revised regularly to improve the efficiency and selectivity of data gathering. The information management system should be flexible and appropriate to the types and quantities of data anticipated. Data management systems should be easy to access and use but should also be secure from unauthorized manipulation or changes. Florida is making a promising attempt to develop and implement a ground water quality information system.

Permanent inventory systems for potential contaminants or sources are helpful in preventing ground water and surface water contamination. One such system is the California Pesticide Registry, which establishes the quantity, location, and timing of the use of chemicals that could have an effect on water quality.

The committee recommends that states consider establishing ongoing inventories of potential contaminating activities and substances. The compounds and activities inventoried should be selected on the basis of potential risks and quantities of use in each state or regional area. They should include not only traditional sources such as industrial discharges, landfills, and underground storage of chemicals and petroleum products, insecticides, herbicides, fungicides and fertilizers but also other polluting substances used in significant quantities in land use practices, such as transportation, septic tank cleaning, drilling or mining operations, and underground injection. Such a system can provide valuable information on quantities and locations of substances being used and their potential for contaminating ground water.

CLASSIFICATION SYSTEMS

"Classification system," as the concept is used in this report, is a comprehensive system with which to classify waters for differential ground water protection strategies. This system may be applied to ground water in several aquifer units or portions thereof and water basins. This basic policy allows states to apply the appropriate levels of protection to ground water based on present and future uses; e.g., high-water-quality aquifers would receive the highest level of protection, whereas other less critical uses of ground water resources would receive lower levels of priority for protection. Criteria for ground water classification include hydrogeologic characteristics, present quality, current and potential uses, land use, vulnerability to contamination, and depth.

A classification system may satisfy a number of objectives, including the following:

- Provide a focus for limited state resources on protection and restoration of valuable and sensitive aquifers;
- Provide a basis for coordinated management of ground water resources and the activities that potentially have an impact on the resource at all levels of government; and
- Provide guidance for implementation of regulatory and enforcement programs.

Under a classification system, geographic areas, aquifers, or portions of aquifers are identified and placed in different categories, each of which is afforded a different level of protection. This allows the state to be mapped according to the designated classification criteria. The nature of the classification system that can be developed is dependent to a great degree on the types of data available. This approach supports the development of cost-effective management strategies and protection techniques.

Connecticut

A number of political, institutional, data, and regulatory framework needs must be satisfied before a successful state ground water classification program can be implemented. In 1980, when Connecticut initiated its program, those factors were in place. The political will to control ground water pollution in Connecticut was demonstrated on May 1, 1967, when the recommendations of a 100-member bipartisan task force were passed as Connecticut's Clean Water Act. Although the primary focus of the act was on Connecticut's surface waters, the law included provisions for the protection of ground water. It is these provisions that provided the underpinnings for the ground water classification system in use today. In the 1960s and 1970s, untreated sewage and industrial wastes were the most visible pollution and received the most attention. However, ground water enforcement actions were initiated as well. The major problems were unrelated to untreated waste waters, salts, nitrates, and solvents. This enforcement activity level and attendant publicity was partially responsible for increasing public awareness of the potential for ground water contamination and the need for effective control.

The hydrogeology of Connecticut is another factor that has an impact on ground water management. Connecticut is generally underlain by a rather shallow bedrock with water wells in the alluvial aquifers commonly running less than 100 feet in depth. Thus, the major ground water basins are nearly synonymous with major surface water drainage basins. Connecticut's ground water management system recognizes the intimate relationship between ground water and surface water. This relationship is also understood by the public, which has made the acceptance of a ground water classification system easier since a surface water classification system was already in existence.

Another factor in developing and maintaining public interest in ground water protection is that approximately one third of Connecticut's population relies on ground water for their water supply source. Twenty percent rely on individual household wells for drinking water without any benefit of routine water monitoring.

The factors described above were instrumental in providing the environment in which ground and surface water management could be integrated into a single consistent program. It was believed that a statewide strategy would be the most effective because each of the 169 towns in the state could not provide or protect ground water to meet their supply demands nor could they provide for proper waste water disposal. However, local planning, zoning, and enforcement were determined necessary for proper implementation of a ground water management system. A system of ground water quality standards and criteria (classification) paralleling the surface water quality standards was determined by the state to be workable and effective. It could be developed in a timely manner, would require public involvement, and could assure consistency in all federal, state, and local permit actions.

In September 1980, Connecticut adopted "Water Quality Standards and Criteria," which included the classification system. The system is based on use standards rather than quality standards. The state took this approach for the following reasons:

- Flexibility to react to changing drinking water standards and new types of pollutants;
- Ability to prohibit certain discharges and land use practices in select areas;
- Ease of enforcement and simplifications of monitoring;
- Encouragement of consideration into local zoning and aquifer protection programs; and
- Consistency with statutory goals.

The resultant policy and use standards are

> . . . to restore or maintain the quality of the ground water to a quality consistent with its use for drinking without treatment. In keeping with this policy, all ground water shall be restored to the extent possible to a quality consistent with Class GA [see Table 4.3]. However, restoration of ground water to Class GA shall not be sought when:
>
> - The ground water is in a zone of influence of a permitted discharge.
> - The ground water is designated as Class GB; unless there is demonstrated need to restore ground water to a Class GA designation or where it can be demonstrated to the commissioner that restoration to Class GA can be reasonably achieved.
> - The ground water is designated Class GC.

Table 4.3 describes the classifications, uses, and discharges allowed.

Converting the management policies and procedures into a working program required classification of all the state's ground water into one of the four use classes established. This required the detailed analysis of the considerable available natural resource data; water supply and waste disposal practices and land use information; application of the state water policy statements; and extensive public workshops, meetings, and hearings. The

TABLE 4.3 Ground Water Classifications, Uses, and Discharges Allowed

Class	Resource use	Compatible discharges
GAA	Public and private drinking water supplies without treatment.	Restricted to waste waters of human or animal origin and other minor cooling and clean water discharges.
GA	Private drinking water supplies without treatment.	Restricted to waste waters of predominately human, animal, or natural origin that pose no threat to untreated drinking water supplies.
GB	May not be suitable for potable use unless treated because of existing or past land uses.	All the above plus it may be suitable for receiving certain treated industrial waste waters when the soils are an integral part of the treatment system. The intent is to allow the soil to be part of the treatment system for easily biodegradable organics and also function as a filtration process for inert solids. Such discharges shall not cause degradation of ground waters that could preclude its future use for drinking without treatment.
GC	May be suitable for certain waste disposal practices owing to past land use of hydrogeological conditions that render these ground waters more suitable for receiving permitted discharges than development for public or private water supply. Downgradient surface water quality classification.	All the above plus other industrial waste water discharges that do not result in surface water quality degradation below established classification goals. The intent is to allow the soil to be part of the treatment process.

SOURCE: State of Connecticut, 1982.

state has completed the classification process that resulted in four major basins being designated (the Thames River basin, the Connecticut River basin, the Central Connecticut Coastal River basin or Quinnipiac basin, and the Housatonic River basin). The adoption and approval of the first three basins occurred in the 1981–1983 term period. The approval of the classification map and program for the Housatonic River basin took considerably longer because of local opposition and was finally completed in 1985.

Figure 4.2 shows a portion of the Water Quality Classification map for southeastern Connecticut. The completed ground water classification maps now serve as a comprehensive statewide blueprint that provides a general

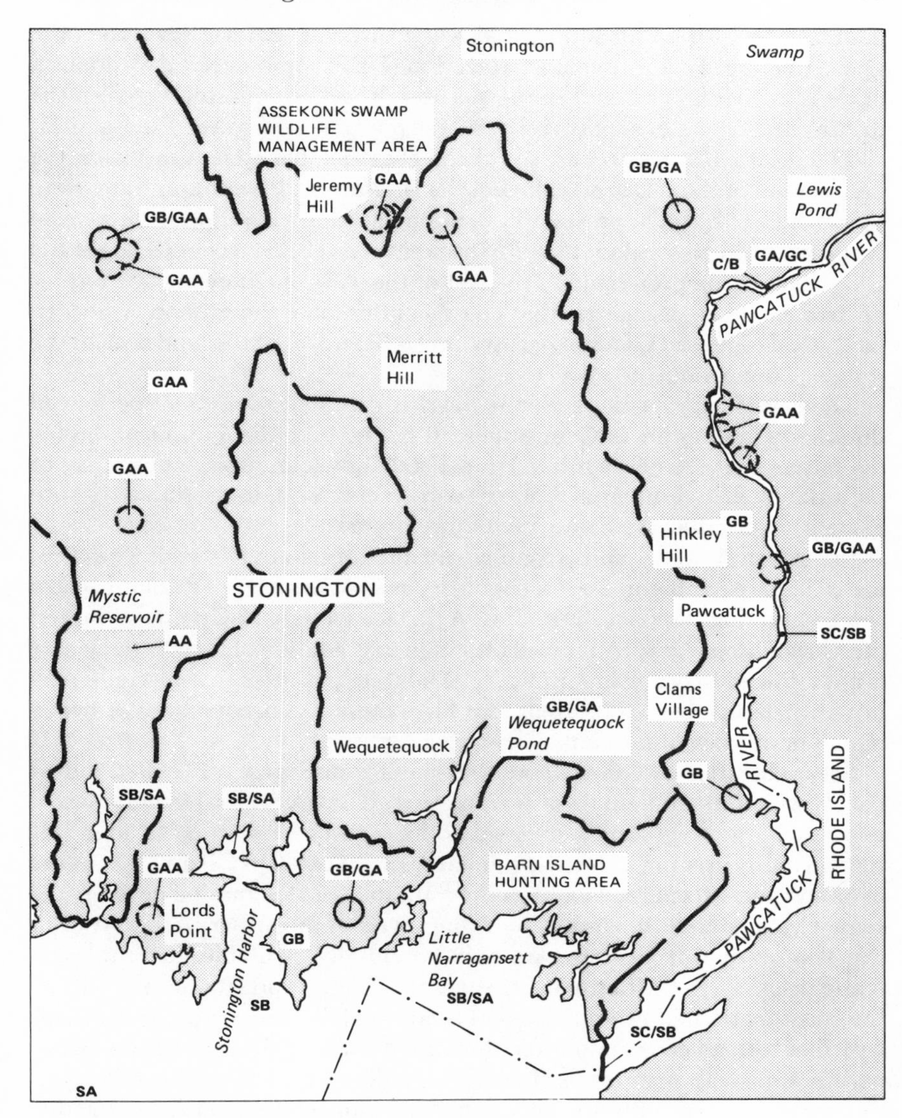

FIGURE 4.2 Examples of water quality classifications on map of southeastern Connecticut. SOURCE: From map of Water Quality Classifications for the Thames, Southeast Coast, Pawcatuck River Basins. State of Connecticut, 1985.

planning tool for water supply and waste disposal. Before 1980 when the Department of Environmental Protection (DEP) commenced the adoption of the classification system, there was literally no relationship between the drilling of new water supply wells and the placement of waste disposal sites.

The most controversial aspect of the classification system has been the designation of the GC areas. Within the Housatonic River basin, they were perceived by the local press and the public as sites where hazardous waste facilities could be located. Despite the controversy, it is recognized that the GC sites provide protection to the rest of the state of Connecticut from industrial or other discharges that could contaminate the ground water. In fact, the designated GC sites occupy only 0.3 percent of the land area of the state of Connecticut.

The classification system affects the siting and location of all permitted discharges in the state of Connecticut. It has no functional implications for discharges that are not regulated under state law, e.g., nonpoint sources of pollution such as residential septic systems or chemical storage tanks that are currently not subject to a state regulatory program.

In general, in GAA and GA areas, no discharges are allowed. These are areas that are at present used or should be useful as sources of drinking water supply and that discharge into Class A surface waters. In GB areas, limited industrial discharges are allowed although degradation of the ground water is not allowed to occur intentionally. The state indicates that it would permit only discharges in GB areas that have minute concentrations of heavy metals or other contaminants.

Areas classified as GC are potential sites for discharges of treated industrial wastes and municipal wastes. In general, any industrial facility with a new discharge to ground water would have to be sited in a GC area, although many existing industrial discharges are located in GB areas. Because of opposition within the Housatonic basin, the GC designation developed a somewhat different meaning from that given in any other basins. Within the Thames basin, GC sites are only potentially available for municipal waste disposal. In the basins other than the Thames and Housatonic basins, the classification allows the disposal of industrial or municipal waste after treatment. In all cases, potential dischargers must go through the necessary permit approvals prior to discharge. According to the state, the classification system affects only the siting, not the control requirements, for discharges. The Department of Environmental Protection (DEP) requires that industrial or municipal dischargers and hazardous waste treatment, storage, and disposal facilities meet applicable Resource Conservation and Recovery Act (RCRA), Best Available Technology (BAT), and state performance or design, and monitoring requirements in all (including GC) areas. However, the classification system may be used to influence the type of remedial action that is taken to clean up or control existing uncontrolled hazardous waste sites or landfills.

The classification system has been used to close landfills. On the maps, certain areas are designated GB/GA, which means that industrial discharges currently exist there, but it is the policy of the state to convert that classified area into a drinking water source where future discharges may be prohibited.

Beginning in 1979 and with an emphasis on the 1981–1982 period, two years of effort went into the preparation of the classification maps for the four basins. This work included doing waste discharge source inventories. Connecticut had the advantage of starting with a good USGS data base.

During this process, the DEP staff put together a number of detailed maps for subbasins for the four major watersheds. These maps include geologic overlay maps showing the location of silt and stratified drift soil deposits (in the GC classification system—DEP looked for fine-grained stratified drift materials next to Class B surface waters), water table elevation maps, depths of bedrock maps, maps depicting the location of major waste sources in water supply wells, and maps designating natural areas, parks, fishing areas, and endangered species habitat sites, for example. The quantity of geologic, hydrologic, water quality, and land use data that the DEP staff has collected and analyzed in preparing the classification maps is extremely impressive. In the decade preceding the adoption of Connecticut's classifications for ground water, the DEP Natural Resources Center had compiled extensive inventories of natural resource information. The most significant element was the compilation of geologic information that described the stratified drift/glacial till distribution and the water table information. Long-term well-drilling reports provided the information on water table elevation, and a 10-year cooperative program with the Hartford office of the USGS provided the geologic information. These data provided the foundation for developing a classification system based on hydrogeologic conditions.

Even with this mass of technical data to support the mapping, refinements in specific areas may be necessary. One industrial firm spent three years and considerable resources to have an area reclassified so that it could continue industrial discharging even while upgrading its waste treatment system. The DEP staff indicates that it will be able to keep the classification system up to date with 1½ to 2 person-years of staff effort annually. Others representing industrial interests indicated that it would take a lot more personnel to keep the classification system working efficiently considering the data needs associated with classifying or reclassifying particular sites.

The DEP staff takes the position that the classification system has been adopted pursuant to statutory requirement, rather than as part of rules and regulations. In Connecticut, the Legislature must approve regulations. As a practical matter, since the classification system reflects the use of an important resource adopted pursuant to statute, it reduces the incentive for industrial firms to challenge it on the grounds that it has not been appropriately adopted as a regulation.

As previously stated, the classification system as it has been designed in Connecticut has a large data base and provides an enormously powerful tool for protecting ground water, the full implications of which have not yet been realized. According to the DEP staff, it has been used to redirect the priorities of the enforcement program. The DEP staff has redirected enforcement efforts toward areas designated as GB/GA on the basin maps, i.e., areas where existing industrial discharges must be phased out or upgraded to comply with the ground water classification. It is evident that, since the GC areas occupy only 0.3 percent of the land area of the state, the adoption of the classification system means that Connecticut is moving rapidly away from dependence on land disposal of hazardous, industrial, and solid wastes.

Classification Use by Industry and Local Units of Government

Industry and local government representatives interviewed by committee members indicated that the classification system had become a powerful tool for industrial and local planning. DEP has prepared a report entitled *Protecting Connecticut's Ground Water: A Guide to Ground Water Protection for Local Officials*. The DEP has one person on its staff who attempts to work closely with local officials to adopt and implement the ground water classification system. The guide shows how the classification system can be used by local planners to control nonpoint sources of pollution through zoning or other enforcement mechanisms not available to DEP. Clearly, it would be a challenge for one staff person to carry out this local coordination function efficiently with all towns in Connecticut. Everyone with whom the committee members spoke agreed that DEP needed more resources to work with the towns and that the towns should be encouraged to revise their ordinances and master plans to protect ground water in accordance with the classification system.

The town of Southington, which has had a problem with ground water contamination owing to a solvents-recovery facility resulting in the closing of some water supply wells, hired a consulting firm to prepare hydrogeologic maps based on the classification system for the town. Southington is using the same classification system as the state has adopted. They will use it for aquifer protection and as a basis for a land acquisition program. Southington modified the state classification map with a more detailed data base, which its consulting firm was able to develop. Other towns will undoubtedly realize that the classification system has an enormous value for local planning.

Colorado

Using survey results and evaluations of other state programs, the Colorado Department of Health (CDH), between August 1983 and September 1984, developed a list of alternative ground water protection goals and a way to achieve the goal selected (CDH 1984). The alternatives considered for a goal were as follows:

1. Maintain existing beneficial uses.
2. Maintain existing and potential beneficial uses.
3. Maintain existing quality.
4. Utilize selective nondegradation of existing quality for certain pollutants.
5. Allow limited degradation of existing quality on a case-specific basis.
6. Formulate a flexible framework to achieve diverse purposes.
7. Continue with no common direction or purpose.

Each goal was evaluated on the basis of three criteria: (1) protection of public health, (2) prevention or control of contamination, and (3) ensuing utilization of ground water. An ad hoc advisory committee of industrial, governmental, and CDH representatives selected goal 6.

In May 1984, the Colorado Water Quality Control Commission (CWQCC) made a final statement of their goal: "The goal of the Water Quality Control Commission is to provide the maximum beneficial use of ground water resources, while assuring safety of the users by preventing or controlling those activities which have the potential to impair existing or future beneficial uses of ground water or to adversely affect the public health."

Regulatory Options for Implementing and Achieving the Goal

Six options for implementing the CWQCC goal were evaluated by the CDH and advisory committees. These were as follows:

1. Formulate classifications and numeric standards of ground water bodies.
2. Require regulations to control waste discharge.
3. Require site-specific permits with effluent limitations.
4. Institute site-specific permits with adequate design criteria.

5. Require generic nondegradation or antidegradation regulation applying statewide.
6. Continue with incomplete ground water regulatory programs.

Four criteria were used to evaluate these options: (1) the ability to achieve the goal, (2) coordination with other regulatory agencies and/or programs to minimize duplication of effort, (3) consideration of economic reasonableness of implementation, and (4) consideration of the amount of data necessary to implement the goal.

Implementation Option

The option selected is a form of classification scheme. Ground water quality is classified on the basis of total dissolved solids (TDS). This differs from classification approaches in some states in that neither physical boundaries (e.g., aquifer systems, recharge areas, geologic strata) nor ground water uses are delineated. The Colorado approach is a broad water quality classification tied to beneficial use and defined by a water quality analysis for TDS in parts per million (ppm).

There are three beneficial-use categories:

Use Category 1: Ground waters with a TDS range of less than 3000 ppm are suitable for all beneficial uses including public and private water supplies, irrigation, livestock and wildlife watering, and commercial and industrial purposes.

Use Category 2: Ground waters with a TDS value of 3,000 to 10,000 ppm are suitable, but not ideally, for most beneficial uses.

Use Category 3: Ground water with a TDS value greater than 10,000 ppm would be generally unsuitable for most beneficial uses except for some commercial and industrial uses.

If a site-specific sample of ground water is found to be in Use Category 1, based on TDS value, then the ground water is protected to keep the TDS in that range of value. Primary and secondary drinking water quality standards have been developed for Use Category 1, with absolute limits for each selected contaminant that could impair use. For Use Category 2, only primary standards apply. No standards apply to Use Category 3. Action for ground water in Use Category 3 is primarily to monitor, not control, water quality.

Both ground water use and protection are under the jurisdiction of the CWQCC. The regulatory framework was written by staff members of the CDH, but the CWQCC has the authority to adopt and enforce the program.

Relationship to Existing Regulations

Colorado's Memorandum of Agreement requires agencies that regulate sources or activities not covered under the New Water Quality Act to have regulations "essentially equivalent" to CDH requirements. All other regulations must result in the performance standard element of the program. Performance standards for sources and activities covered in the program are being developed to ensure protection of ground water quality in Use Categories 1 and 2.

Appendix C to this report contains the following information relating to regulations protecting Colorado's ground water:

1. Activities with an impact on ground water to be controlled by the new program, ranked in order of potential for causing contamination (11 activities).
2. Activities that may have an impact on ground water but for which adequate information is not yet available to evaluate specific impacts (12 activities).
3. Activities that may have an impact on ground water but that are specifically covered by other state or federal regulatory programs (18 activities).
4. Activities exempted from coverage under the ground water quality control program (3 activities). These activities still must meet performance standards to the extent that water rights are not materially injured.

Application of the Program

Site-specific studies are required by potential dischargers in all cases. Factors considered are existing uses of ground water, potential uses, geologic conditions, water quality, and other factors as appropriate.

If a facility is considered to be a potential source of contamination, the owners are requested by CDH to monitor for ground water quality impact. If monitoring confirms contamination, CDH requires the owners or operators to propose and implement a remedial action plan.

Owners or operators of a proposed facility must make a site review. Initially, the TDS is used to determine Use Category designation. Then, on a site-specific basis, the Use Category designation is refined to consider other factors such as potential ground water use, geologic conditions, and other conditions. This site review then becomes the basis for CDH action.

New Jersey

The cornerstone of New Jersey's ground water management and protection activities is the state ground water discharge permit program pursuant to the New Jersey Clean Water Act of 1976. A simplistic aquifer classification system based on total dissolved solids (TDS) has been put into place as one of the factors that is considered in the setting of effluent limitations. The classification system has not been used as a proactive aquifer mapping tool, with the exception of the Central Pine Barrens area, rather it is used on a site-specific basis as one of the factors that determine permit limits. The data gathering is simple, as the classifications are based either on TDS levels and are easily determined for a given site, or on geographical boundaries for the Pine Barrens area of the state.

In actual practice, the New Jersey classification system has not been useful. Of the approximately 500 ground water discharge permits issued since 1981 by the New Jersey Department of Environmental Protection (DEP), virtually all have been issued in class GW2 or the fresh water (TDS of 500 ppm or less) classification category. The DEP is currently evaluating the potential usefulness of various more-sophisticated classification systems and plans to put into place a different system by early 1987.

The present classification system was developed in 1981 under the provisions of New Jersey DEP rule-making authority, where rules concerning ground water quality standards were promulgated. The rules addressed policy with respect to protection and enhancement of ground water resources, use classification, quality criteria, and the designated uses of ground waters of the state pursuant to the New Jersey Water Pollution Control Act and the Water Quality Planning Act. These policy developments recognized the close association of ground water and surface water.

The policy calls for nondegradation except where the state chooses to allow lower water quality as the result of necessary and justifiable economic or social development. In no event, however, may degradation of water quality interfere with or become injurious to existing designated uses. Also, no degradation is allowed in the Central Pine Barrens geographical area. In cases where water quality does not meet the listed criteria for a particular use classification owing primarily to human activities, it is the policy of the DEP to restore the quality to the minimum levels defined for the use classification. When water quality does not conform for natural causes, natural water quality characteristics shall prevail.

The ground water classification system has the following categories:

- Class GW1 ground water, which applies only to the Central Pine Barrens, shall be suitable for potable water supply, agricultural water supply, continual replenishment of surface waters to maintain the existing quantity and high quality of the surface waters in the Central Pine Barrens, and other reasonable uses.

- Class GW2 ground water having a natural TDS concentration of 500 mg/L or less shall be suitable for potable, industrial, or agricultural water supply, after conventional water treatment (for hardness, pH, iron, manganese, and chlorination) where necessary for the continual replenishment of surface waters to maintain the quantity and quality of the surface waters of the state and other reasonable uses.
- Class GW3 ground water having a natural TDS concentration between 500 and 10,000 mg/L shall be suitable for conversion to fresh potable waters or other reasonable beneficial uses.
- Class GW4 ground water having a natural TDS concentration in excess of 10,000 mg/L shall be suitable for any reasonable beneficial uses. Effluent limits and quality criteria will be determined on a case-by-case basis for these waters.

Quality criteria for Classes GW1 through GW3 are given in Table 4.4.

TABLE **4.4** Quality Criteria for Class GW1 Through GW3

Pollutant, substance, or chemical	Ground water quality criteria
Class GW1: Ground Water Quality Criteria for the Central Pine Barrens (Class GW1 applies only to Central Pine Barrens)	
1. Aldrin/dieldrin	1. 0.003 μg/L
2. Arsenic and compounds	2. 0.05 mg/L
3. Barium	3. 1.0 mg/L
4. Benzidine	4. 0.0001 mg/L
5. Cadmium	5. Natural background
6. Chromium (hexavalent) and compounds	6. Natural background
7. Cyanide	7. 0.2 mg/L
8. DDT and metabolites	8. 0.001 μg/L
9. Endrin	9. 0.004 μg/L
10. Lead and compounds	10. 0.05 mg/L
11. Mercury and compounds	11. 0.002 mg/L
12. Nitrate-nitrogen	12. 2.0 mg/L
13. Phenol	13. 0.3 mg/L
14. Polychlorinated biphenyls	14. 0.001 μg/L
15. Radionuclides	15. Prevailing regulations adopted by EPA pursuant to sections 1412, 1415, and 1450 of the Public Health Services Act as amended by the Safe Drinking Water Act (PL 93-523)

(*Continued*)

TABLE 4.4—*Continued*

Pollutant, substance, or chemical	Ground water quality criteria
16. Selenium and compounds	16. Natural background
17. Silver and compounds	17. 0.05 mg/L
18. Toxaphene	18. 0.005 μg/L
19. Ammonia	19. 0.5 mg/L
20. BOD (5-day)	20. 3 mg/L
21. Chloride	21. 10 mg/L
22. Coliform bacteria	22. (a) by membrane filtration, not to exceed 4 per 100 mL in more than one sample when less than 20 are examined per month; or (b) by fermentation tube, with a standard 10-mL portion, not to be present in three or more portions in more than one sample when less than 20 are examined per month, or (c) prevailing criteria adopted pursuant to the federal Safe Drinking Water Act (PL 93-523)
23. Color	23. None noticeable
24. Copper	24. 1.0 mg/L
25. Fluoride	25. 2.0 mg/L
26. Foaming agents	26. 0.5 mg/L
27. Iron	27. 0.3 mg/L
28. Manganese	28. 0.05 mg/L
29. Odor and taste	29. None noticeable
30. Oil and grease and petroleum hydrocarbons	30. None noticeable
31. pH (standard units)	31. 4.2–5.8
32. Phosphate, total	32. 0.7 mg/L
33. Sodium	33. 10 mg/L
34. Sulfate	34. 15 mg/L
35. Total dissolved solids	35. 100 mg/L
36. Zinc and compounds	36. 5 mg/L

Class GW2: Ground Water Quality Criteria Statewide Where the Total Dissolved Solids (TDS, natural background) Concentration Is Less Than or Equal to 500 mg/L

Primary Standards/Toxic Pollutants

1. Aldrin/dieldrin	1. 0.003 μg/L
2. Arsenic and compounds	2. 0.05 mg/L
3. Barium	3. 1.0 mg/L
4. Benzidine	4. 0.0001 mg/L
5. Cadmium and compounds	5. 0.01 mg/L
6. Chromium (hexavalent) and compounds	6. 0.05 mg/L
7. Cyanide	7. 0.2 mg/L
8. DDT and metabolites	8. 0.001 μg/L

TABLE 4.4—*Continued*

Pollutant, substance, or chemical	Ground water quality criteria
9. Endrin	9. 0.004 μg/L
10. Lead and compounds	10. 0.05 mg/L
11. Mercury and compounds	11. 0.002 mg/L
12. Nitrate-nitrogen	12. 10 mg/L
13. Phenol	13. 3.5 mg/L
14. Polychlorinated biphenyls	14. 0.001 μg/L
15. Radionuclides	15. Prevailing regulations adopted by EPA pursuant to sections 1412, 1415, and 1450 of the Public Health Services Act as amended by the Safe Drinking Water Act (PL 93-523)
16. Selenium and compounds	16. 0.01 mg/L
17. Silver and compounds	17. 0.05 mg/L
18. Toxaphene	18. 0.005 μg/L
	Secondary Standards
19. Ammonia	19. 0.5 mg/L
20. Chloride	20. 250 mg/L
21. Coliform bacteria	21. (a) by membrane filtration, not to exceed 4 per 100 mL in more than one sample when less than 20 are examined per month, or (b) by fermentation tube, with a standard 10-mL portion, not to be present in three or more portions in more than one sample when less than 20 are examined per month, or (c) prevailing criteria adopted pursuant to the federal Safe Drinking Water Act (PL 93-523)
22. Color	22. None noticeable
23. Copper	23. 1.0 mg/L
24. Fluoride	24. 2.0 mg/L
25. Foaming agents	25. 0.5 mg/L
26. Iron	26. 0.3 mg/L
27. Manganese	27. 0.05 mg/L
28. Odor and taste	28. None noticeable
29. Oil and grease and petroleum hydrocarbons	29. None noticeable
30. pH (standard units)	30. 5–9
31. Phenol	31. 0.3 mg/L
32. Sodium	32. 50 mg/L
33. Sulfate	33. 250 mg/L
34. Total dissolved solids	34. 500 mg/L
35. Zinc and compounds	35. 5 mg/L

(*Continued*)

TABLE 4.4—*Continued*

Pollutant, substance, or chemical	Ground water quality criteria
Class GW3: Ground Water Quality Criteria Statewide Where the Total Dissolved Solids (TDS, natural background) Concentration Is Between 500 mg/L and 10,000 mg/L	
Primary Statewide/Toxic Pollutants	
1. Aldrin/dieldrin	1. 0.003 μg/L
2. Arsenic and compounds	2. 0.05 mg/L
3. Barium	3. 1.0 mg/L
4. Benzidine	4. 0.0001 mg/L
5. Cadmium and compounds	5. 0.01 mg/L
6. Chromium (hexavalent) and compounds	6. 0.05 mg/L
7. Cyanide	7. 0.2 mg/L
8. DDT and metabolites	8. 0.001 μg/L
9. Endrin	9. 0.004 μg/L
10. Lead and compounds	10. 0.05 mg/L
11. Mercury and compounds	11. 0.002 mg/L
12. Nitrate-nitrogen	12. 10 mg/L
13. Phenol	13. 3.5 mg/L
14. Polychlorinated biphenyls	14. 0.001 μg/L
15. Radionuclides	15. Prevailing regulations adopted by EPA pursuant to sections 1412, 1415, and 1450 of the Public Health Services Act as amended by the Safe Drinking Water Act (PL 93-523)
16. Selenium and compounds	16. 0.01 mg/L
17. Silver and compounds	17. 0.05 mg/L
18. Toxaphene	18. 0.005 μg/L
Secondary Standards	
19. Ammonia	19. 0.5 mg/L
20. Chloride	20. Natural background
21. Coliform bacteria	21. (a) by membrane filtration, not to exceed 4 per 100 mL in more than one sample when less than 20 are examined per month, or (b) by fermentation tube, with a standard 10-mL portion, not to be present in three or more portions in more than one sample when less than 20 are examined per month, or (c) prevailing criteria adopted pursuant to the federal Safe Drinking Water Act (PL 93-523)
22. Color	22. None noticeable
23. Copper	23. 1.0 mg/L
24. Fluoride	24. 2.0 mg/L
25. Foaming agents	25. 0.5 mg/L

TABLE 4.4—*Continued*

Pollutant, substance, or chemical	Ground water quality criteria
26. Iron	26. 0.3 mg/L
27. Manganese	27. 0.05 mg/L
28. Odor and taste	28. None noticeable
29. Oil and grease and petroleum hydrocarbons	29. None noticeable
30. pH (standard units)	30. 5–9
31. Phenol	31. 0.3 mg/L
32. Sodium	32. Natural background
33. Sulfate	33. Natural background
34. Total dissolved solids	34. Natural background
35. Zinc and compounds	35. 5 mg/L

SOURCE: Regulations Implementing the New Jersey Water Pollution Control Act, Chapter 14A, Subchapter 6, Ground Water Quality Standards—April 1985.

Conclusions and Recommendations

A comprehensive classification system such as that used in Connecticut can be an effective tool for optimizing ground water protection efforts. Maps prepared on the basis of a classification system can be used to guide activities such as the development of standards for water supply, land use management, source controls, and remedial action. By directing the location of potential sources of pollutants away from critical areas, classification can also reduce the cost and controversy associated with case-by-case siting of facilities. In addition, a mechanism for coordination between state and local governments is provided.

Where mapping is not feasible, because of divided authority or data limitations, classification can still provide guidance, especially during permitting and enforcement procedures. However, its usefulness in this case is more reactive than helpful as a planning tool. This is true of the classification systems in Massachusetts and in New Jersey outside the Central Pine Barrens zone, where all fresh ground water is essentially considered to be one class (i.e., drinking water).

The committee recommends that states consider classifying their ground water in conjunction with a mapping program that specifically identifies critical areas and resources for special protection. If the data are not sufficient, state and local efforts should be directed at collection of the necessary information to provide for classification and mapping in a phased approach. The lack of sufficient data should not necessarily preclude the development of a classification

system. The criteria should be adopted through a public process and state and local efforts directed at the establishment of criteria for classification and gathering of necessary data to prepare maps to implement the system.

Comprehensive classification programs depend on adequate hydrogeological information to be effective. Development of the Connecticut classification system was, in large part, due to the existence of historical hydrogeological information produced by the USGS.

The committee recommends that the USGS expand its efforts to produce hydrogeological information to support state and local ground and surface water protection programs.

GROUND WATER QUALITY STANDARDS

Statewide standards for ambient ground water quality are intended to establish upper limits of concentration of designated pollutants in ground water consistent with the use of those waters for beneficial uses. The states that the committee investigated have all discovered evidence of contamination of ground water by synthetic organic chemical compounds in recent years. A significant issue that they face is what ambient standard for individual organic contaminants they should set to protect ground water for beneficial uses that might include drinking water, irrigation, and ecological protection. The task of developing those standards is complex. State and local governments do not generally have the ability or resources to develop scientifically based standards. For this reason, most states have looked to EPA to perform this function or to provide technical information that the states can use in setting ambient standards for these pollutants. In turn, the states can use ambient standards as one basis for regulatory and enforcement action to limit discharges of those pollutants from point and nonpoint sources into ground water.

Under the federal Safe Drinking Water Act (SDWA) of 1974, EPA is supposed to prepare and promulgate regulations or standards for constituents of health concern in public water supplies. The SDWA calls on EPA to set two different kinds of standards for water used for human consumption: recommended maximum contaminant levels (RMCLs) and maximum contaminant levels (MCLs). The RMCLs represent maximum concentrations of pollutants based solely on health concerns. Under the SDWA, EPA may not enforce these limits, they are primarily informational and represent long-term goals. By contrast, the MCLs are enforceable. If a public water supply exceeds a MCL for a pollutant, the purveyor is required to take action to reduce concentrations of that pollutant below the MCL. In late 1984, EPA proposed RMCLs and MCLs for many additional organic and inorganic chemicals. The information that EPA presented in its public notice on this proposal should be of immense assistance to the states.

Many states have used the RMCLs and MCLs as a basis for setting ambient ground water quality standards. However, EPA has adopted RMCLs and MCLs for only a limited number of constituents, including very few organic chemical compounds. The states find themselves having to deal with an increasing number of compounds for which standards have not been developed. Where MCLs and RMCLs are not available, states must develop their own standards based on their own analysis or based on Suggested No Adverse Response Levels (SNARLs), which are guidelines suggested by EPA for a limited number of organic chemicals. Some states, such as New York and New Jersey, have set ambient standards for ground water and standards specifically applicable to drinking water.

Another issue closely related to ground water quality standards is soil quality standards. Water that percolates through contaminated soils can leach contaminants and carry them to ground water. Therefore, the question of what are acceptable levels of contaminants in soil becomes important. None of the programs examined by the committee included specific soil quality standards. However, each of the states and localities indicated that the issue was an important one and was difficult to resolve. The principal difficulty is that the potential impact of soil contaminants on ground water quality is highly site specific. Therefore, uniform standards for soil quality would generally be inappropriate.

Wisconsin

In 1984, Wisconsin enacted a law (Wisconsin Act 410, see Appendix D), that required that each regulatory agency identify all substances already detected in ground water or substances that have a reasonable probability of entering the ground water that result from activities that the agencies regulate. Ground water quality protection standards are then to be developed for each of those substances. In Wisconsin, these standards are based on a "two-tiered" approach; for each substance identified, an "enforcement standard" and a "preventive action limit" (PAL) will be set.

Standards for those substances that are determined to be a public health concern (e.g., certain organic compounds, heavy metals) are to be recommended by the Wisconsin Department of Health and Social Services, while the Wisconsin Department of Natural Resources (DNR) will recommend standards for those substances determined to be a public welfare concern (chloride, turbidity, and TDS, for example). Standards for each substance are then adopted by rule by the DNR and apply to all regulated activities and agencies.

Currently, ground water quality standards have been proposed in Wisconsin for over 50 substances for which federal standards have previously been established by the EPA or recommended by the National Research Council (NRC, 1977–1983). Standards for other substances will be developed and some previous standards modified as new risk information becomes available. However, the DNR reportedly intends to establish standards that may or may not be consistent with federally determined standards. There is significant concern, therefore, that the federal standards will result in pressure for the state to enforce the less stringent federal standards.

In Wisconsin's two-tiered system, the adopted ground water quality standards become enforcement standards. Enforcement standards define when violations of ground water quality standards have occurred and apply to all state-regulated activities that have an impact on ground water quality. When a substance is detected in ground water in concentrations equal to or greater than its enforcement standard, the source is subject to immediate enforcement action. The appropriate regulatory agency must prohibit continuation of the activity from which the substances came, unless it can be demonstrated that an alternative response will achieve compliance with the enforcement standard.

The preventive action limit (PAL) represents a ground water quality standard that is a lower concentration of the substance than the enforcement standard. PALs are intended to function as warning levels and as standards in facility design. These limits have been established by the state for each of the substances with enforcement standards. For the substances of public health concern (volatile organics and heavy metals, for example), PALs are set at 20 percent of the enforcement standard except where the substance is reported to be carcinogenic, mutagenic, or teratogenic. For these substances, the PAL is set at 10 percent of the enforcement standard. PALs for the public-welfare-related substances (chloride, turbidity, and TDS, for example) are set at 50 percent of their enforcement standards.

These more stringent limits must be used in design standards for facilities (e.g., landfills) and in management practices (e.g., pesticide use regulations) so that contamination up to or greater than the enforcement standard is prevented. Regulatory agencies are required to review their existing design code regulations to assure that they conform to the PALs to the extent technically and economically feasible. The PAL is also intended to serve as a "trigger" for regulatory response. Exceeding a PAL creates the possibility that some regulatory response may be necessary. In that case, the regulatory agency is required by law to evaluate the situation and take action necessary to maintain the concentration of the substances at the PAL or at the lowest concentration technically feasible. The agency may prohibit continuation of the activity that is the source of the problem. However, to do so the agency must meet specific statutory requirements. PALs are intended to provide regulatory agencies with time to take preventive measures to ensure that the enforcement standard is not reached or exceeded.

Preventive action limits apply everywhere ground water is monitored, even within the property boundaries of a facility. Enforcement standards will apply at

- any point of ground water use;
- any point beyond the property boundary of a regulated facility; and,
- any point outside the subsurface attenuation zone provided by what are known as "design management zones" (DMZs) for regulated activities.

DMZs are subsurface zones extending horizontally a specified distance from the particular source and vertically downward from the water table through the entire saturated zone.

Kansas

Kansas has used ground water quality standards for chloride and specific conductance since the 1950s to protect ground water for several beneficial uses. Levels of these parameters were established on the basis of considerations of effects on human health and livestock. For drinking water use, concentrations of chloride must be less than 500 mg/L, and water under 5,000 mg/L is protected for use by livestock. For ambient ground water with concentrations of less than 5,000 mg/L chloride or 10,000 mg/L TDS, polluting activities are not allowed. With respect to organic compounds, Kansas uses a two-tiered approach to standards similar to that in Wisconsin. Action levels have been adopted for 29 primary volatile organic compounds based on maximum contaminant levels (MCLs) established by the EPA. In addition, the state has established Notice Levels (which are provided to both private and municipal water suppliers) corresponding to concentrations 1/100 (0.01 times) the MCLs to provide opportunity to react to a problem before MCLs are exceeded.

Florida

Florida has adopted a unique continuation of numerical and descriptive standards for its four classes of ground water, two classes of which are for drinking water. The standards provide the basis for enforcement action. The state employs federal drinking water limits where available. EPA (non-enforceable) Health Advisories and SNARLS are used for additional organic compounds. A cancer risk of one in one million is the standard for carcinogens. Where the state deems the available data to be inadequate to set a standard, the standard is the detection limit. This is accomplished through a narrative standard that calls for all Class I and II ground waters to be "free from" all toxic chemicals.

New Jersey

New Jersey's classification system illustrates the application of a combination of health and environmentally based water quality standards to ground water. For most of the state, health-based standards for drinking water apply. However, in the Central Pine Barrens a fragile surface water ecosystem is fed by ground water discharge. Ground water standards were therefore required for this region that would protect surface water quality. Natural background ground water quality was determined through a monitoring program. Background levels of phosphate, nitrate, pH, and ammonia were then established as ambient quality standards for this class GW-1. These standards apply in addition to drinking water standards, which protect private well users in the Pine Barrens region. This is a unique example in which ambient standards were set to prevent any degradation of background water quality (see "Land Use Controls" section later in this chapter).

Connecticut

The Connecticut classification system, described earlier in this chapter, contrasts with the above systems in that only narrative water quality standards were adopted. The standards describe the beneficial use of each class and the general types of discharges that would be allowed. Numerical compliance standards are established for each facility in the course of the enforcement of the discharge permit program. Available criteria for drinking water quality, including federal advisories and local standards, are taken into consideration at that time. The state believes that the use of narrative standards streamlines the regulatory process by eliminating public standard-setting procedures. In addition, it provides the state with greater flexibility in its regulatory program.

California

Ground water quality standards in California are set by the nine Regional Water Quality Control Boards, under statewide policies set by the State Water Resources Control Board (SWRCB). These standards are established to protect the beneficial uses designated in the Basin Plans. Specifically, the regional boards set waste discharge requirements for all surface dischargers and all discharges to ground water if there is a threat to water quality (fresh water injection to maintain a hydraulic barrier against sea water would not constitute a threat, for example, whereas reclaimed waste water used for the same purpose would). These discharge requirements are contained in a permit that places effluent standards on what can be discharged and that sets a timetable for any required treatment upgrading. The permit usually establishes a schedule of monitoring and reporting requirements, the reports to be made under penalty of perjury. Enforcement powers provided to the state and regional boards include the capability to take polluters to court and to levy fines directly (up to $5000 per day). As a matter of course, cleanup orders are issued first, in an attempt to deal with the problem—fines are used as a last resort. Water quality standards adopted under the Basin Plans are reviewed by regulation at three-year intervals and updated as necessary. This process is the subject of public hearings.

Rule-making to promulgate standards varies widely. For example, in California, as stated previously, ambient ground water quality standards are set by the nine regional boards acting under the aegis of the State Water Resources Control Board. On the other hand, drinking water standards are established and monitored by the State Department of Health Services (DOHS). Where possible, the DOHS uses federal standards as their own. However, because national drinking water standard development has lagged behind the actual measurement of synthetic trace organics in wells, the DOHS has resorted to the use of "action levels" based on the judgment of its professional staff, and criteria published by the EPA, the National Research Council, or others. Whereas the ambient and drinking water standards are attended by formal rule-making procedures, the setting of action levels may occur on an ad hoc basis.

Conclusions and Recommendations

Water quality standards are set at various levels of government and for different purposes. Federal drinking water standards apply to all public drinking water supplies. Additional standards for drinking water at the point of use may also be adopted by states.

Depending on their policy, states may apply numerical standards directly to ground water. These may be designed to protect drinking water, other beneficial uses, and critical ecological systems. They may also be used to define nondegradation of high-quality waters. In some cases a safety factor is built in, where the standard is set at a fraction of the enforcement limit. This allows for future growth and the uncertainty associated with ground water protection technology.

In view of the complexity of setting standards to assist in protection of ground water, the committee recommends the following:

- **EPA should proceed expeditiously with promulgating the RMCLs and MCLs that it has recently proposed, and EPA should propose and promulgate RMCLs and MCLs for all other inorganic and organic chemical compounds commonly found in ground water.**
- **EPA should also continue to provide technical information to states about the organic chemicals in ground water for which it has not promulgated RMCLs or MCLs.**
- The application of numerical standards to ground water is a matter of state policy, and there is no single approach that would be appropriate on a national basis. **Ambient ground water standards should be based on the individual state's adopted goals and objectives. These may include protection of beneficial uses other than drinking water, nondegradation, and protection of ecological systems.**
- Wisconsin is one state reviewed by the committee that has developed a two-tiered set of standards designed to limit degradation of ground water and require action by polluters. **In setting standards the states should consider a multitiered standard-setting approach that can be used to justify nondegradation of high-quality ground water and to protect public health.**
- **In addition, the committee recommends that EPA provide states with a central permanent source of technical research and standard-setting criteria. However, EPA should have the capability to establish overriding standards when states establish inconsistent standards hindering effective long-term prevention of ground water degradation.**

CONTROL OF CONTAMINATION SOURCES

Management of Hazardous Materials and Waste

Proper hazardous waste disposal capability available to generators who cannot handle their waste on site is of critical concern for the protection of ground water quality. It is especially needed to minimize the illegal dumping that has been found in all parts of the country. Many states either do not have such facilities or have prohibited any establishment of hazardous waste facilities within their borders.

A comprehensive program to manage hazardous materials and hazardous waste is essential for protection of ground water. An effective state or local program should include the following:

- A process for defining hazardous materials and hazardous waste and identifying sources of generation, storage and handling, transportation, treatment, and disposal.

- A regulatory and enforcement process to ensure safe management of hazardous materials and hazardous waste during production, use, handling, storage, treatment, and disposal.
- Incentives for reduction of hazardous waste production.
- A policy for encouragement of safe hazardous waste treatment and disposal by individual home owners and industry within the state or local jurisdiction.
- A process to control and restrict the location of hazardous material and hazardous waste activities to protect significant sources of ground water.
- A process for training government, industry, and the public on effective strategies for safe management of hazardous materials and hazardous waste.
- Strategies to encourage all segments of society, including individuals, to manage hazardous materials and hazardous waste properly.

A number of the state and local government programs examined by the committee have some interesting approaches to hazardous materials and waste management that are of particular importance to prevention of ground water contamination.

Florida

The 1983 Water Quality Assurance Act prohibits land disposal of hazardous waste. It also prohibits the Florida Department of Environmental Regulation (DER) from permitting new underground injection control (UIC) wells that inject hazardous waste; old UIC wells are grandfathered. A serious problem is that a good portion of the hazardous waste that is not allowed to be disposed of through the land disposal methods in the state is transported to Alabama, South Carolina, and other states. Florida has no approved commercial hazardous waste facilities that render waste nonhazardous through incineration, neutralization, or other processes.

The state is conducting a comprehensive hazardous waste assessment. It is designed to identify all industrial, commercial, and institutional generators of hazardous waste, irrespective of their size. The hazardous waste survey may provide Florida with information useful in determining its long-term hazardous waste disposal strategy.

Florida recognizes that ground water contamination can result from improper disposal of solvents, pesticides, and other organic chemicals by large numbers of individuals. It has conducted some "amnesty days," when individuals or firms can bring hazardous materials to central collection areas. All permit applicants must consider the impact of their facilities on all environmental media, not just ground water.

Dade County, Florida

As part of an aggressive program to protect wellfields from ground water contamination, Dade County, Florida, has established a strong hazardous waste management program. Dade County has a much broader definition of what constitutes hazardous waste than does the state or EPA. At present, Dade County has listed in excess of 900 chemicals as hazardous. Pursuant to the State Hazardous Waste Inventory Program, Dade County contracted (with funding from DER and the Regional Planning Council) with a consultant to identify potential hazardous waste sources. Through its survey, it has identified 8000 generators.

The county has a program that requires a permit from any firm that handles, generates, or disposes of hazardous waste. Its IW-5 (Industrial Waste Regulation) permitting program regulates all those who handle or generate 55 gallons or more of hazardous waste per year. A newly proposed IW-6 program would require permits for all nonresidential industrial septic tanks. The program seems to be keeping the regulated community aware of county programs.

The county has broad legal authority to regulate every industrial discharge and has effective enforcement capability. It has legal authority to obtain emergency injunctions in County or Circuit Court, with relief including punitive and compensatory damages. It has brought about 100 enforcement cases in County or Circuit Court and has won all cases. The Department of Environmental Resources Management (DERM) has 15 inspectors in 15 districts. Each inspector controls a district and is therefore able to be familiar with virtually every business and every significant source. These inspectors, with help from fish and game wardens, have found illegal hazardous waste dumpers. DERM also has six people assigned to enforce underground storage regulations, eight permit reviewers, and four specialty inspectors. The program is financed by user fees. The county intends to conduct inspections once every year on every regulated source.

New Jersey

New Jersey has a unique approach to cleanup and control of hazardous waste that places the responsibility for cleanup on industry before property can be transferred to a new owner. The Environmental Cleanup Responsibility Act of 1983 (see "Land Use Controls" section below) requires certain industrial establishments to obtain state certification that there has been no discharge of hazardous substances or waste on the property or that any such discharge has been cleaned up according to department-approved procedures before the property can be sold. The New Jersey Department of Environmental Protection (DEP) has the power to void any sale and fine establishments not in compliance with the law. Industrial establishments with certain standard industrial code (SIC) categories that are involved in the generation, manufacture, refining, transportation, storage, handling, or disposal of hazardous substances or hazardous waste above or below ground are subject to the statute. The program has been controversial and costly to industry but very effective because industry must take greater responsibility for management of hazardous substances and waste, which has resulted in cleanup of contaminated ground water and prevention of future contamination. In the first 18 months of the program, the DEP processed 1000 applications. Two hundred required cleanup before sale, and 93 cleanups have been completed. Lending institutions and title insurance companies are taking a more active role in seeing that industry is meeting its environmental responsibilities. New Jersey has a much broader definition of hazardous waste than does EPA, regulates small-quantity as well as large-quantity generators, and has an aggressive enforcement program.

The New Jersey Pinelands Commission has strong land use and regulatory authority over most economic activities, including industrial and municipal waste generating and disposal activities, in the New Jersey Pine Barrens, an area that covers 22 percent of the state's land. No storage or disposal of hazardous materials or hazardous waste (other than gasoline or household quantities) by industrial, commercial, or governmental entities is allowed in the Pine Barrens. All land disposal of municipal solid waste is prohibited unless there is no feasible alternative outside the Pine Barrens. All existing landfills must be closed by 1990 unless a waiver is obtained because there is no feasible alternative outside the Pine Barrens.

Arizona

In Arizona, the state legislature finally addressed the problem because the nearest disposal site for the major urban and industrial centers was over 400 miles away in Nevada. A law was enacted establishing a state hazardous waste facility, to be financed, constructed, operated, and maintained by a private contractor.

The choice of the site was actually a political one, with the legislature making the selection. It was not the top candidate from the list of suitable sites presented to the legislature by the Department of Health Services (DHS) but was one of the first two alternates. The one-square-mile site, originally on U.S. Bureau of Land Management (BLM) property, was acquired by the state but is still surrounded by BLM land.

Acquisition of the site was a protracted process since it was the first time the federal government had been asked for land for such a purpose. The heads of the Department of the Interior, EPA, and the Council on Environmental Quality became involved in the question of whether an environmental impact statement was necessary on the transfer of the land, who would pay for it, and who would have the lead responsibility. In all, it took 34 months from the original request for purchase of the land by Arizona to the actual sale. Considerable effort was also expended to meet with the citizens in the area during the siting process to address or mitigate their concerns. Additional citizen involvement activities will be necessary as the facility is constructed and becomes operational. The private contractor selected by DHS, which has the oversight responsibility for the facility, proposed a $14 million high-temperature facility capable of incinerating PCBs and other hazardous waste. Negotiations with the contractor were completed in November 1985.

The facility, which will recycle as much of the waste as possible, is scheduled to open in 1986. It is recognized that this type of high-tech facility will handle waste from other states since the waste generated in Arizona alone would not make the operation economically feasible.

California

Hazardous waste management in California is overseen primarily by the Department of Health Services (DOHS). The department's powers derive originally from California's Hazardous Waste Control Act of 1972, and more recently from its delegation to administer the Federal Resource Conservation and Recovery Act (RCRA) of 1976.

Within the DOHS, the Toxic Substances Control Division (TSCD) is primarily responsible for implementing and enforcing these acts. The RCRA program has approximately 320 employees, with headquarters in Sacramento, and three field offices located in Berkeley, Fresno, and Los Angeles. Surveillance and enforcement are carried out from the field offices. DOHS field office personnel are augmented through memoranda of understanding with 10 counties and a number of cities. For example, in Southern California, there are 20 DOHS surveillance professionals in the field, augmented by 150 professional staff members of local health departments and by other environmental professionals. It is noted that many of the latter are heavily committed, however, to enforcing the underground container program discussed below. In addition, TSCD coordinates its field operations with those of other involved agencies, including local district attorneys, law enforcement agencies, CALTRANS, and the California Highway Patrol.

The TSCD oversees three broad categories of hazardous waste entities: first, the treatment, storage, and disposal facilities; second, the waste haulers; third, the generators. The dimensions of the problem are large and not fully known. There are an estimated 70,000 hazardous waste generators in California, widely ranging in size. Of these, some 22,000 as of 1984 were within TSCD's inventory for surveillance. TSCD has registered and permitted some 1300 treatment, storage, and disposal facilities, as well as 1100 haulers. Of these, the division stated that it inspected most of the treatment, storage, and disposal facilities during fiscal 1984–1985, plus more than 1900 of the generators. Critics of the division believe that many of these inspections are cursory and lack depth (Environmental Defense Fund, Berkeley, personal communication, 1985). On the other hand, some in the regulated community believe that the division and other local enforcement groups are unreasonably harsh.

During 1983–1984, some 3½ million tons of hazardous waste were shipped off site for treatment, storage, and disposal. Most such waste eventually is hauled to seven landfill sites located in California. This immense volume of material is tracked by manifest, such that the TSCD typically receives 25,000 manifests each month—12,500 generator copies and 12,500 disposal copies; the federal form is used. A spot check in June 1985 placed the number of manifests at 27,300. A centralized computer system, with distributed terminals (there are four in the Los Angeles office, for instance) is used for entry and retrieval of the data. Field managers indicate that this record-keeping burden is extremely difficult, although of the highest practical importance. Many of the forms are filled out by people who have not filled out such forms previously, and mistakes are made and must be caught and corrected through quality control. In addition, there are lags in reporting of up to a month. Critics of the TSCD believe that the manifest system is inadequate, and incomplete, with missing data (Environmental Defense Fund, Berkeley, personal communication, 1985). Field users and managers, on the other hand, while acknowledging difficulties, feel that the system is extremely useful and is used many times each day by surveillance personnel (TSCD, Los Angeles field office, personal communication, 1985). Surveillance personnel use the retrieval capabilities of the system routinely to identify, for example, how much of waste X from counties Y and Z was hauled during a particular period to disposal site W. The retrieval function is user-friendly and driven by an understandable menu so that the working professional does not have to become a computer expert to use the system.

Inspections and spills identify violations of the law. During 1983–1984, more than 1250 violations were written by TSCD staff members. Of these, 304 had been mitigated as of the fall of 1985, 43 had been cleaned up, and legal action had commenced on 103. TSCD reports levying fines and penalties amounting to more than $850,000 under RCRA.

Field managers are frank to state that the problems they face are immense and that there are critical needs in the prevention area. One of the highest needs cited is the conversion to a treatment and neutralization-based system, rather than a landfill-based system. It is not clear how to reach this solution; meanwhile, landfilling continues for lack of an alternative, subject to increasing numbers of prohibitions. A number of highly toxic wastes have been banned from landfills, such as all halogenated solvents and heavy metals above certain concentrations. These streams, and others, must now be treated prior to disposal. Source reduction is a second obvious need cited by field personnel and would complement a treatment-based system.

Conclusions and Recommendations

The committee finds that land disposal of hazardous waste is a serious source of ground water contamination that requires urgent attention.

Therefore, the committee recommends that, as an essential element of each state's ground water protection program, a plan should be developed for treating, storing, or disposing of hazardous waste within its boundaries. A program for waste minimization should be a key element in the plan. Such a program should also include a siting process for transportation, storage, and disposal facilities, including regional and on-site industrial incinerators. Exportation of hazardous waste, a temporary expedient that generally increases risks associated with transportation and decreases the assured overall level of environmental protection, should be considered or continued only in special circumstances. The federal government should have a role in mediating this decision in case arbitration is needed.

A unique approach to cleanup and control of hazardous waste was found to be New Jersey's Environmental Cleanup Responsibility Act of 1983 (ECRA), which places responsibility for cleanup on industry before sale of property to a new owner. This type of legislation provides an effective prevention as well as remedial pollution control program.

Therefore, the committee recommends that other states should consider adoption of programs comparable to ECRA with broad application to provide incentives for good housekeeping by industrial firms, municipalities, and other significant polluting activities.

Management of Municipal Solid Waste

A comprehensive program to manage solid waste is an essential component of a state program designed to prevent ground water contamination. Sanitary landfills, the most frequently used method of solid waste disposal, and incinerator and resource recovery ash must be properly controlled.

A comprehensive state solid waste program should include the following:

• A regulatory, surveillance, and enforcement program to control the transfer, treatment, and disposal of solid waste including resource recovery facility ash. Such a program should include approval of plans, specification and issuance of permits, leachate and ground water monitoring, and financial responsibility of facility owners for proper waste management;

• Requirement for and assistance to counties or municipalities on short- and long-term solid waste management plans and their implementation;

• Training for generators and disposers, including the public, on recycling, managing, and disposing of solid waste;

• Programs to encourage recycling, resource recovery, and industrial processing that minimize solid waste production;

• A process to control and restrict the location of solid waste disposal facilities to protect significant sources of ground water;

• The ability to monitor compliance and take enforcement actions where necessary.

Solid waste landfills are a major source of ground water contamination. State and local governments are also finding that the capacity of existing municipal landfills is very limited, and they are searching for alternative ways of managing solid waste. New Jersey, New York, and Kansas provide examples of states that are promoting planning for solid waste handling and disposal on a county basis.

In general, these states are encouraging construction of incineration/resource recovery plants. However, state emission and ash disposal standards and monitoring requirements for solid waste resource recovery plants are limited. In addition, few of the state and local governments have assessed and implemented aggressive solid waste recycling programs.

New Jersey

Most New Jersey communities are running out of municipal solid waste landfill capacity. Siting of new landfills is a problem because of their potential to pollute ground water.

The New Jersey Solid Waste Management Act, N.J.S.A. 13:1E-1, requires each county to develop its own solid waste management program by itself or in conjunction with one or more other counties. Another state law establishes a minimum recycling goal for each county of 25 percent with some state tax incentives to promote this goal.

Because of the sensitivity of the one-million-acre Pine Barrens to ground water contamination and ecological disruption, the New Jersey Pinelands Commission in its Comprehensive Management Plan has prescribed stringent controls on land disposal of solid waste in the Pinelands. All landfills in the most sensitive central portion of the Pinelands, known as the Preservation Area, had to cease operating when the plan was adopted in 1981. Other existing landfills must close by 1990, and no new landfills may open unless a solid waste management district demonstrates to the commission that no practical alternative disposal techniques are available in or outside the Pine Barrens.

New York

Long Island, New York City, and many other communities in the state of New York are considering alternatives to land disposal of solid waste. The primary alternative is incineration/resource recovery. New York enacted a bond issue that provides communities with up to 50 percent of the cost of constructing resource recovery facilities.

Nassau and Suffolk counties, Long Island, face a serious solid waste crisis because the major municipal landfills are situated in the central ground water recharge areas of the island, designated as the deep flow recharge area in both the Long Island Section 208 waste water management plan and the draft New York Department of Environmental Conservation (DEC) Long Island Ground Water Management Plan. At least one municipal landfill, the Bethpage landfill, is on the Superfund National Priority List.

An outstanding problem is the siting of ash disposal facilities. Two sites are currently being considered—one in the town of Brookhaven near the existing Brookhaven landfill, which is at the southern border of the deep flow recharge areas of Long Island's main aquifer. The Stony Brook University Marine Science Research Center is also conducting research in using resource recovery ash to build artificial reefs. Whether the state DEC will classify such ash as a hazardous waste subject to state and federal RCRA disposal requirements is still an open question.

The Board of Estimate in New York City has recently approved the city's first resource recovery plant with a proposed site at the Brooklyn Navy Yard. A recent Environmental Defense Fund (1985a) report, *To Burn or Not to Burn: The Economic Advantages of Recycling Over Garbage Incineration for New York City*, indicates that recycling of solid waste in the city is cost effective relative to construction of incineration plants. The city's Department of Sanitation has indicated a willingness to conduct a recycling program on a fully equivalent basis with the Brooklyn Navy Yard incineration project, over and above the recycling that occurs under the state's beverage container deposit law. This law has reduced the volume of solid waste in the state by 5 to 8 percent; in the city it has been reduced by 4 percent.

Disposal of solid waste in Nassau and Suffolk counties is now controlled by the Long Island Land Use Land Fill law, ECL 27.0704 (see Appendix F). That law prohibits the siting of new or expansion of existing solid waste municipal landfills within the deep flow recharge areas. Furthermore, it gradually phases out continued land disposal of municipal solid waste in existing landfills in the deep flow recharge areas and, eventually, anywhere in Nassau and Suffolk counties. The exceptions include land disposal of the ash residue from resource recovery plants and up to 10 percent solid waste bypass.

In view of the prohibitions in the Long Island Land Use Land Fill law, the townships in Nassau and Suffolk counties are under pressure to site and construct resource recovery facilities. The Long Island regional office of DEC has indicated that towns, as a goal, must establish solid waste recycling programs to recycle initially 10 and then 20 percent of solid waste as a condition of state permits for the resource recovery plants.

As in most other states, emission standards applicable to resource recovery plants are in a state of flux. The New York State DEC has recently issued draft operational guidelines for such plants. The draft guidelines, which include combustion efficiency and acid gas controls, propose limited stack testing but no specific emission limits for noncriteria pollutants such as heavy metals (arsenic, beryllium, cadmium, chromium, lead, mercury, nickel), polychlorinated dibenzo-p-dioxins, polychlorinated dibenzofurans, polycyclic aromatic hydrocarbons (BaP, chrysenes), formaldehyde, and polychlorinated biphenyls.

Kansas

The Kansas program for solid and hazardous wastes is unique in two respects. The planning and implementation of solid waste disposal was financed without state aid. New York State was at the other extreme by providing all the funding for counties to study the matter and develop plans. Kansas was the first state to have permitted waste collection and disposal facilities for the entire state in 1976.

The Kansas Solid Waste Management Act, passed in 1970, outlawed open dumps and required a planning committee to prepare a solid waste management plan for each county. The plans were to be completed by 1974. Cities could be included in the county plan or apply for approval of their own plans. By the end of 1976, all counties except two had systems functioning with approved sanitary landfills. One of these two counties hauled waste to an out-of-state site, and the other contracted for disposal in the landfill of a neighbor. Observation wells were installed at landfills in an effort to detect any migration of contaminants into the ground water.

The importance of giving special attention to the disposal of hazardous materials became more apparent as the county planning progressed, and Kansas made an inventory of hazardous materials in 1975. The findings of the inventory and passage of Public Law 94-580, The Resource Conservation and Recovery Act of 1976 (RCRA), led to major modifications of the Kansas act in early 1977 to provide a complete program for the control of hazardous waste.

The Kansas program emphasizes process changes to reduce waste as well as recycling and detoxification. When storage is the only solution, it must be above ground. An intense, lengthy controversy about the state's only approved disposal site has caused its closing, and so hazardous waste must be hauled to sites in other states. Experience to date has shown that in Kansas the major threat to ground water is from petroleum products and chemicals leaking from old storage tanks and pipelines.

Cape Cod, Massachusetts

The Cape Cod Economic Development and Planning Commission and the Cape Cod Township have recognized that the Corps of Engineers' municipal landfills, many of them situated away from developed coastal areas over its prime recharge areas, are major sources of ground water contamination. Cape Cod towns and Barnstable County therefore are exploring the feasibility of participating in the large regional resource recovery plant under consideration in Rochester Township, just off the Cape proper.

Since Massachusetts has a beverage container deposit law, a large portion of the glass, metal, and plastic beverage container component of the Cape's solid waste is now recycled. Otherwise, Cape Cod local governments are not promoting large-scale recycling programs as an alternative to off-Cape resource recovery.

Connecticut

The Connecticut Department of Environmental Protection (DEP) has used their classification system to close landfills, e.g., the landfill in Colchester. Some firms that treat industrial waste need economic long-term solid waste landfill sites. Some of these firms are looking at sites in GC areas (see Table 4.3) designated on the statewide classification maps. Several private firms have looked for expanded landfill capability and have not been able to find it. Connecticut must use these GC sites rather sparingly and press for non-land-based disposal techniques for industrial and municipal waste for its classification system to maintain its integrity.

Under the 1973 State Resource Recovery Authority Act, the state policy has been to move away from land disposal of solid waste. The State Assembly realizes that Connecticut municipalities are running out of space for landfill. A great many landfills have closed, and about 75 of the 169 towns must now transport their solid waste to other towns in the state.

The Connecticut Legislature has enacted a new Solid Waste Management law. Under this law, the individual towns have until 1987 to develop short- and long-term solid waste management plans. The statute establishes a process for the state DEP to implement solid waste management plans in those towns that do not have a DEP-approved plan. The statute is designed to move the towns toward regional resource recovery plants. Connecticut also has a beverage container deposit law.

Florida

Landfills are a major potential source of ground water contamination in Florida. State policy seems to encourage alternatives to land disposal through construction of solid waste incineration/resource recovery plants. Florida law does not encourage source reduction or recycling of such materials in that it has no beverage container legislation at present.

Since so much of the land mass of Florida is important for ground water recharge, the phase-out of land disposal of solid waste has decided benefits for ground water protection. Florida has at least two large resource recovery plants in operation, in Dade County and in Pinellas County. The operator of the Pinellas County plant is an engineer who designed it. The mass burn technology utilized by the Pinellas County resource recovery facility reduces the volume of solid waste by 90 percent; by weight, the reduction is 75 percent. That plant also provides for rear-end ash separation with a removal and subsequent sale of ferrous metals by scrap and industrial firms. Seven to 10 percent of the ash is metals, and the remainder of the ash is deemed not toxic under the state ground water rule. It is now being used as a landfill cover, although the Department of Environmental Regulation may authorize experiments with ash aggregate in road construction.

Each incineration plant in Pinellas County is equipped with an electrostatic precipitator, and emissions must meet a particular opacity and odor standard. The third unit now under construction must satisfy three air quality standards, plus standards for lead, nitric oxides, sulfur dioxide, and carbon monoxide. Resource recovery plants in Florida are subject to Florida's air quality regulations. These regulations do not include emission standards and monitoring requirements for potential resource recovery hazardous air pollutants, such as the dioxins and acid gases. Siting of these plants is regulated under the State Power Plant Siting Act.

Wisconsin

Through its Bureau of Solid Waste Management, the Wisconsin Department of Natural Resources licenses all landfills annually. New landfills must have clay liners and leachate collection systems, and recent ground water protection legislation authorizes the department to require the installation of monitoring systems around existing landfills. Wisconsin also imposes a tax on landfills that have not entered into an agreement to close by 1995. It is designed as an incentive to phase out use of small town landfills that have no monitoring. The department also has a program designed to encourage solid waste recycling with a state recycling coordinator.

Conclusions and Recommendations

The committee finds that both nonhazardous and hazardous solid waste disposed of on land are major contributors to ground water contamination. However, the quantities of waste needing disposal in landfills can be reduced through recycling, incineration, and resource recovery facilities.

The committee recommends that states and communities consider such methods of reducing waste quantities, but only as part of an integrated environmental management program with monitoring requirements, discharge or emission limits, and ambient environmental quality standards for both ground water and air resources that use comparable concepts of risk assessment.

Underground Storage Tanks

Underground tanks are widely used for interim storage of toxic and hazardous materials. They are also used, less commonly, for storage of toxic and hazardous wastes. These tanks, found in and near every population center and transportation corridor, pose a significant and continuing threat to ground water resources.

Some state and local programs designed to protect aquifers from leaking underground storage tanks (LUST) have existed for many years. Nevertheless, incidents of leaking tanks and aquifer damage have increased in number and severity. Mounting concern has resulted in regulatory action by EPA, as well as strengthened and innovative programs in a number of states. State and local programs generally incorporate two parallel strategies: (1) provision for identification and control of existing underground tanks and (2) regulation of the siting, design, construction, and monitoring of new tanks.

Existing Tanks

Registration Prerequisite to the effective control of underground storage tanks is the development and maintenance of an inventory of existing tanks. Fourteen of the 15 towns on Cape Cod (Barnstable County, Massachusetts) have adopted local by-laws requiring registration of all underground gasoline, fuel, or chemical storage systems. Most by-laws, which were adapted from model regulations prepared by the Cape Cod Planning and Economic Development Commission, exclude residential fuel oil tanks from this requirement. Kansas has required registration of underground tanks since 1975.

Although New York State has a comprehensive bulk storage law, Nassau and Suffolk counties (Long Island) have adopted county health ordinances that are more restrictive regarding underground storage of petroleum products and other chemicals. Suffolk County's ordinance (Article 7, Suffolk County Sanitary Code) requires owners/operators of underground tanks to obtain permits, renewable every five years, from the Department of Health Services. Tanks smaller than 1100 gallons that are used to store fuel oil, diesel oil, kerosene, or lubricating oil prior to use (not resale) are exempted.

Dade County, Florida, goes a step further by requiring owners of existing chemical tanks to obtain permits, which are renewable annually. The county reserves the right at its discretion to add conditions to the permit at time of renewal, such as provision of sampling points and monitoring wells.

Inspection One approach used to identify existing tanks that are leaking is through a periodic inventory reconciliation and physical inspection program. The Cape Cod towns require annual or semiannual inventory verification and conduct annual physical inspections and tests of all registered tanks at least 15 years of age or older. Suffolk County, Long Island, inspects existing tanks on a biennial basis for tanks 20 or more years old and every 4 years for tanks less than 10 years old. The county reports that the incidence of leaking tanks has fallen from 15 percent of all tanks inspected in 1981 to 2 percent in 1984.

Prohibitions Kansas prohibits underground or at-grade storage of toxic and hazardous waste materials everywhere in the state. Dade County, Florida, does not permit tanks without secondary containment in designated wellfield areas and prohibits use of unprotected steel tanks (without cathodic protection, or secondary containment) everywhere in the county.

The Cape Cod towns require tanks not in conformance with current standards to be removed by a set date. Suffolk County, Long Island, requires all nonconforming tanks to be removed by 1990 and prohibits reuse or resale of nonconforming tanks. When removed from service, they must be filled with sand or concrete in place or removed and perforated so as to be unfit for reuse and sold as junk.

California is now implementing an underground container registration and regulation program. The program was initiated in 1983 and 1984 by AB 1362 (Sher) and AB 2013 (Cortese). Because of the scope of the effort, state personnel are augmented by personnel from cities and counties, who share with the state the responsibility for permitting, inspection, and enforcement. Some 165,000 tanks have been registered to date, and the process is considered nearing completion. The tanks include gasoline and other fuels, chemical storage, sump wastes, and others. The program is self-financing through fees levied on the registrants.

New Construction

The Cape Cod towns require new tanks to be constructed of noncorrosive materials (e.g., fiberglass) or to provide corrosion protection (coatings or cathodic protection, for example). New fuel storage facilities at residential sites (single family and two family) may not be located underground; existing underground tanks at these sites must be phased out within 30 years of original installation.

Dade County, Florida, requires all new installations to meet specific design standards. Tanks constructed in designated wellfield areas must be provided with secondary containment and some means of detecting leakage into the zone between the two containment layers.

New facilities in Suffolk County, Long Island, designed for materials only slightly soluble in water, and with specific gravities less than one, must be constructed of noncorrosive materials or otherwise protected from corrosion by coatings, cathodic protection. All other facilities must be double-walled, with sampling access between primary and secondary containments. All new facilities must incorporate either overfill protection devices or a product-tight containment capable of intercepting and retaining overfill spills.

In addition, under its Article 7, the Suffolk County Department of Health Services (see Appendix E) restricts the siting of industrial operations that use hazardous materials. In general, Article 7 imposes those restrictions within the deep flow recharge areas of Suffolk County, i.e., within Zones III and V as designated in the Long Island Comprehensive Waste Treatment Management Plan (1978), Section 703, paragraph P, which lists the toxic or hazardous materials covered by the Article 7 restrictions. Suffolk County's underground storage tank program is implemented by the County Department of

Health Services. A total of 18 full-time-equivalent positions are allocated to the permitting, testing, design approval, and enforcement activities.

The Connecticut DEP has the legal authority to regulate underground fuel tanks, and these regulations are currently under review. The law authorizes DEP to delegate these powers to local units of government, thus DEP is delegating these programs to fire marshalls. It must now approve these underground fuel tanks for safety. Further, the state does not generally regulate chemical storage tanks.

California's law has similar provisions for new construction, which apply statewide. Implementation has been slowed by difficulty in promulgating specific guidelines, and deadlines have been reset a number of times. Annual pressure testing or installation of a continuous leak-monitoring system will be required. Companies may be required to pull up existing chemical tanks and double-line them. Because gasoline tanks are so numerous, only new ones must be double-lined. The average gas station will pay about $3000 to 4000 to meet state rules, and then pay an annual cost of about $1500 for monitoring or testing. In addition to the monitoring requirements, spill reporting, closure requirements, and variance procedures are included. Counties must regulate under the law, issue permits to implement the state regulations by stated dates (which have been changing), inspect tanks every three years, and handle small spills. The initial costs of compliance have been estimated at $1,146 million.

Conclusions and Recommendations

Many states and localities, such as Long Island, New York, and Dade County, Florida, have effective underground storage tank control programs that will reduce ground water contamination from this source.

Therefore, all states should consider developing a comprehensive program for monitoring and inspecting chemical and petroleum product storage tanks with stringent design standards for all new tanks and a requirement for monitoring, testing, and upgrading existing tanks in important recharge areas.

Nonpoint Source Contaminants

There are several groups of nonpoint sources of ground water degradation, including chemicals used in the production of plants and animals, the improper use of solvents or other synthetic organic compounds, on-site domestic waste disposal, e.g., septic tanks, mining, and highway maintenance. Chemicals are used by agriculture to control pests (above and below ground), promote growth, and control weeds. They are also used in large quantities for the same purposes to maintain urban landscapes. Synthetic organic chemicals such as TCE are used to clean suburban and rural septic tank waste disposal systems. Salt applied to roadways, waste oils, brake lining materials, road oils, and urban runoff also contribute to ground water degradation.

These sources are more difficult to control than point source pollution since contamination from them occurs on a widespread basis. Control requires regulations to restrict the use of hazardous chemicals or provide incentives to minimize discharge of these contaminants.

Agricultural Pesticides Applied to Land

Agricultural pesticides applied to the land are distinct from other types of ground water contaminants in that they include "economic poisons," which are designed to be toxic and are applied to large areas of the environment. Besides ground water contamination, these chemicals can adversely affect workers, air and surface water quality, and consumers of food products.

Pesticide contamination is a serious problem in many areas of the country. In Long Island, New York, almost 2000 private drinking water wells have been contaminated with aldicarb (trade name Temik), an insecticide and nematicide. About 1000 of these wells have aldicarb concentrations that exceed the New York water quality standard of 7 ppb (parts per billion). Nine other pesticides have been detected in Long Island wells. Since 1979 almost 2500 wells in California have been found to be contaminated with dibromochloropropane (DBCP), including at least 1473 wells that exceed the California Department of Health Services standard of 1 ppb. Aldicarb has also been found in 24 wells in Del Norte County, California. Ground water contamination from EDB,1,2-D and simazine has been traced to lawful agriculture use in California. There has also been ground water contamination from improper disposal at pesticide manufacturing plants.

In a sampling of 70 public wells in Iowa, atrazine was found in 24 wells (34.2 percent) of 14 water supplies (35.96 percent). Monitoring also detected cyanazine, alachlor, metolachlor, and fonofos. In contrast to the situation in Long Island and California, all the concentrations detected in Iowa were below federal and Iowa water quality standards. It is possible that many more pesticides make their way into the ground water, but the difficulty and expense of detection keeps them hidden.

The extraordinary advances in detection equipment within the last decade now allow discovery of contamination at levels below most current health advisory levels. But the widespread application of chemicals makes detection expensive and difficult to plan. Thus, low levels of nonpoint contaminants may go undetected. The control of pesticide contamination of ground water can be accomplished by avoiding the use of toxic pesticides in areas where they will be persistent and migrate or where the ground water is very shallow. However, to accomplish this goal while assuring vigorous agricultural production requires the following:

- Sufficient data on the environmental and toxicological characteristics of each of the many chemicals in use.
- Extensive information on pesticide use patterns.
- Adequate monitoring to enable early detection of any unforeseen problems.
- The development of alternative methods, including integrated pest management, changes in application timing and procedures, and more environmentally benign pesticides, which will promote cost-effective crop management that reduces the threat to ground water.

The following sections discuss these items and describe how some of the states have used these methods.

Large Numbers of Chemicals in Use Tens of thousands of pesticide formulations, containing about 600 different active ingredients, are registered with EPA. About 150 active ingredients account for 90 percent or more of the pounds of pesticides applied in most states. The usage patterns, the toxicological effects, and the chemical propensity to migrate into ground water vary tremendously among active ingredients. Although the number of pesticides that pose a threat to ground water is probably less than 100, it is a huge task for EPA and the state agencies to evaluate which pesticides can safely be used in which geographical locations and in which crop production systems. Federal funding for this task has not been adequate, and, as a result, regulatory decisions are often made without adequate information. It is clearly not cost-effective for each state to develop basic data on the toxicity and environmental fate of pesticides.

Data on Patterns of Use Because the use of pesticides is so diffuse and their fate uncertain, it is very important to know where and when potential contaminants have been applied. The only state that collects and maintains detailed information on pesticide usage is California. The data have been collected for only about 80 pesticides that have been designated as "restricted," primarily on the basis of toxicity rather than leachability. Unfortunately, the "restricted" list does not currently include many pesticides (especially herbicides) that have the chemical potential to leach into ground water. Pending legislation (AB 2021, 8/28/85) in California does set up a usage data collection program for leachable pesticides.

States that do not have pesticide usage information can attempt to estimate spatial distribution of use from areal sales data and cropping patterns. However, such methods are usually not very accurate and do not give the fine resolution necessary to identify the location of a contaminant. The USDA also maintains a data base of pesticide usage, but the information is not adequate for determining the location of potential ground water contamination.

In California, a restricted pesticide can only be applied by a professionally licensed pest control advisor (PCA) or a grower who has received a special license for applying a specific pesticide. In order to apply a restricted pesticide, the applicator must report in advance the pesticide brand, name and formulation, the location (to the nearest square mile), the dosage, the crop, and the time of application. The applicator cannot obtain the permit more than one week in advance of the application. These data are then keypunched and entered into a computer. The California Department of Food and Agriculture (CDFA) summarizes the data in publicly available reports that give the number of applications, the number of pounds, and the total number of acres treated in California.

The pesticide use data have been essential in determining where to monitor for the presence of specific pesticides, and have both reduced the cost for analytical tests and increased the speed with which new contamination events could be determined.

The University of California at Davis can produce computer-generated maps of each county that show to the nearest square mile the location and amount of use for each restricted pesticide. This information has been used to determine which wells should be monitored and which chemicals the analysis should be designed to detect.

Although the principle of a data usage base is laudable, there have been significant problems with the implementation of this approach in California. There have been some inaccuracies in the data, and the total given in the CDFA annual summary use report does not equal the amount of pesticides sold as reported in the CDFA annual report on pesticides sold. A report by the Auditor General of California (1984) also criticized CDFA because the department did "not specify how these [use data] reports should be used in achieving goals." The auditor general's report also complained that the CDFA does not know the extent of inaccuracies in the summary use reports because the department does not know (a) the number of use reports that are not returned by the district offices and (b) the number of use reports rejected by the computer. There have also been inaccuracies in converting from gallons of pesticide to pounds of active ingredients. Apparently as a result of this problem, the total usage given in CDFA's annual summary report does not equal the totals given in CDFA's report on pesticides sold.

These difficulties do not negate the benefits of California's pesticide use data base system, but they do point out the importance of ensuring accurate and complete reporting, which should be an essential element of any new state program in this area.

Predicting Which Pesticides May Cause Ground Water Contamination

Whether a pesticide will reach ground water depends upon a number of interactions between the soil, the water in the soil, and the pesticide. These interactions determine how long the pesticide survives in the soil (persistence) and how far it moves (mobility). Persistence and mobility are determined by many factors, including the following: water solubility, volatility, soil sorption, and degradation (including differing reactions to light, water, and soil microorganisms). The significance of these chemical characteristics depends upon the local soil conditions (including pH and percent organic matter), temperature, moisture, precipitation, and ground water flow patterns.

To assess the likelihood that a pesticide will reach ground water, it is important to know its chemical characteristics and the local soil conditions. For example, aldicarb is very water soluble; thus, it is not surprising to find it in ground water overlain by sandy, quickly draining soils. On the other hand, a water-insoluble material such as trifluralin, which generally absorbs very strongly to soil, would appear in ground water only after a long time if at all. Information of this type can be used to predict which locations and usage patterns are likely to result in pesticide movement into soils. Use of mobile, persistent pesticides can be prohibited in areas where analysis of the pesticide's characteristics compared with information (soil types, pH, drainage pattern, location of aquifer, depth to ground water, for example) about the location indicates that a potentially serious contamination could result from the pesticide. The use of aldicarb in some counties with sandy soils or shallow water tables has been prohibited in California and New York. Unfortunately, this action was not taken until aldicarb was detected in a number of wells.

Because of the large number of interacting factors, it is frequently difficult to decide which pesticides will contaminate ground water just by evaluating individual quantitative measures of chemical and soil characteristics. For example, low soil mobility can be offset by persistence. The interval between the application of a pesticide and the occurrence of the next rainfall (or irrigation) also may have a major impact on the pesticide contamination reaching ground water. A rainfall soon after an application will leach a pesticide before it has a chance to degrade. Also, the variation in soil conditions can mean that chemicals that pose a potentially serious problem in some conditions may be used safely in other situations.

The variability in hydrogeology and soil conditions in Florida has resulted in significant regional differences in ground water contamination from aldicarb. In the northwest potato-growing area in Florida, very little aldicarb has been detected despite shallow ground water and a long history of aldicarb use. However, subsurface drainage to surface water in this area does contain substantial aldicarb concentrations. It is believed that the hydrogeology and alkaline conditions in this region cause rapid degradation of aldicarb. In contrast, in the central citrus-growing region aldicarb has been detected in a number of areas. In the central region, very sandy soils, acidic soil conditions, low soil organic content, and shallow water tables promote aldicarb's persistence and its ability to reach ground water (Holden, 1985).

In order to examine the interaction between environmental fate characteristics of pesticides and soil conditions, mathematical models have been developed that attempt to predict the movement of a pesticide as a function of its chemical characteristics and usage patterns and as a function of local soil and climatic conditions (Carsel et al., 1985; Jury et al., 1983). CDFA plans to use such models to identify which pesticides should be prohibited in which areas within the state.

Registration of Pesticides and Data Gaps

To be sold as a pesticide, a material must be granted a "registration" by the EPA. Most states also have their own state boards that review the data and decide whether to grant a registration for the state. Local officials may also have authority to prevent registration in a certain area because they believe that the material poses a threat to ground water. Aldicarb, for example, is registered for use in New York and California, but is prohibited from use in one county in California and on Long Island.

The Environmental Protection Agency requires data to be submitted by the manufacturer to determine if the pesticide meets health and safety standards. Federal law prohibits EPA from registering a pesticide unless the applicant submits data to prove the product will not cause "unreasonable adverse effects on man and the environment." (Reference U.S. Code, Section 136a(c)(5)(D).) Unfortunately, the "adverse effects" standard was not enforced until 1972. Because of slow progress in reviewing and updating the registrations of older pesticides, many maintain federal registration without meeting existing data requirements or satisfying the toxicological and environmental standards that newer products must meet in order to gain registrations.

The Federal Environmental Pesticides Control Act of 1972 (an amendment to the Federal Insecticide, Fungicide and Rodenticide Control Act—FIFRA) originally required EPA to review and reregister more than 50,000 formulated pesticide products by October 1976. This deadline was later amended to be 1977 and further amended to remove the deadline entirely. EPA has not been able to complete this process in part because the scientific information available for many of the pesticides is inadequate to determine if they are safe. At the current time, there is an incomplete set of toxicological data for many pesticides currently in wide use and relatively few products—unfortunately, no more than one quarter are fully in compliance with the agency's existing data requirements promulgated in Part 158 of the Code of Federal Regulations. (U.S. EPA, 1981; Committee on Policy Research Management, 1985; and House Committee on Agriculture, 1983.) Environmental data that can be used to estimate the likelihood that pesticides will reach ground water under certain soil and precipitation/irrigation conditions are even more scarce. A major data call-in program was undertaken about 1980, and a wide range of data pertinent to assessing the propensity of a pesticide to reach ground water was requested from manufacturers of 84 pesticides.

In 1984, two researchers from the Department of Environmental Toxicology at the University of California, Davis, reviewed the toxicological and environmental data in the files of the California Department of Agriculture for eight pesticides, all of which were heavily used in California. Of the eight pesticides, it was found that five did not have adequate data to determine the pesticide's potential to contaminate ground water. Missing data included photo-degradation, soil mobility, hydrolysis, and octanol-water partition coefficient. One researcher found that three of the pesticides did not have adequate health data. Missing information included acute oral toxicity, long-term chronic studies, as well as studies on metabolic, teratogenic, mutagenic, or reproductive effects. It was not clear from their report if the data were missing because EPA never received them or because California never requested the information from EPA. Without the environmental data, it is very difficult to make an objective assessment of the potential of a particular pesticide to leach and of the severity of the resulting ground water contamination.

In an attempt to eliminate data gaps in pesticide information, California recently enacted legislation (Assembly Bill 2021, 8/28/85) that will eventually cancel registration in California for any pesticide for which a complete set of environmental fate data is not available. The environmental data required includes water solubility, octanol-water partition coefficient, Henry's Law constant, soil sorption coefficient, soil metabolism, photolysis, hydrolysis, vapor pressure, and field dissipation. The dates of cancellation range from 1986 to 1989 and depend upon the method of pesticide application and the CDFA director's willingness to grant an extension. No new pesticides can be registered after January 1986 unless all of the data requirements are met.

Based on the environmental fate data, the CDFA director is required by AB 2021 to establish a list of economic poisons that have the potential to pollute ground water. The selection of pesticides for this list must be based on numerical values that are at least as stringent as EPA's flagging criteria. Pesticides placed on the Ground Water Protection List will be regulated similarly to restricted pesticides. Applicators must report to the commissioner within a week after they use a pesticide on the Ground Water Protection List. All those pesticides on the Ground Water Protection List will be put on the list of restricted-use pesticides. Hence, applicators must obtain permits a week before use and provide detailed information on the location of application. This information then will be incorporated into the detailed use data base discussed earlier.

Monitoring

An important element of ground water protection is monitoring. It is expensive to chemically analyze water samples to detect low levels of pesticide concentrations. The cost of conducting good monitoring programs is increased by the large number of pesticides in use and the large area over which they may be used. It is not feasible to sample every ground water aquifer for every pesticide registered for use. However, if information on usage patterns, chemical characteristics, and hydrogeologic conditions is used, a cost-effective program can be developed. Monitoring costs could be reduced, and a broader understanding of potential problems could be achieved by the development of improved, multiresidue analytic screens for pesticides in water. Such development should be possible at reasonable costs within a few years. Comparable multiresidue analytic methods detecting up to 125 active ingredients in a single test are routinely used by the Food and Drug Administration (FDA) in its food residue testing programs.

Because of the high spatial variability in pesticide use and accompanying variability in ground water contamination, a monitoring program is most effective if pesticide use data, hydrogeologic conditions, and environmental fate information are considered in determining the sampling program. Environmental fate data can be used to determine which pesticides should receive high priority for monitoring because of their propensity for leaching. The use data coupled with hydrogeologic information can be analyzed to determine where samples should be collected. Toxicological data may also be used to assign priorities to pesticides for inclusion in a sampling program. Hydrogeologic and soil information can be used to set geographic priorities for sampling (Cohen and Bowes, 1984).

Given the large number of pesticides and the geographic range of their use, the use data in California have been essential for developing cost-effective ground water monitoring programs. For example, in 1977, DBCP was discovered in the ground water near a California plant that produced the pesticide. The state instituted a ground water monitoring program to look for the nematicide DBCP and used the Pesticide Use Data Base to determine where sampling should be done. A single crop application of DBCP has resulted in contamination in some areas. Therefore, it was important that the location of all previous usage be documented. By 1984, about 2500 well samples containing DBCP contamination had been found. Many of these contamination events are in small, private wells, and the high spatial resolution of the use data was important in determining which wells would be sampled. Subsequent determination that other nematicides (EDB,1,2-D and 1,3-D) were also environmental threats led to additional well monitoring for those chemicals, which was also guided by the spatially detailed pesticide use data.

The economic value of the Pesticide Use Data Base was again illustrated when, in 1983, the California Legislature passed AB 1803, a law requiring widespread monitoring for a range of potential pollutants. Although the bill initially included an appropriation for $4 million, the law was signed with no funds appropriated for the study. With regard to pesticides, the Department of Health Services determined 40 "priority" pesticides that were most likely to be serious pollutants. They initially considered 880 large water systems (200-plus hookups). However, to require 880 systems to sample for 40 pesticides would be excessively expensive and unwarranted. Since no special appropriation was made for this study, it was important that the cost for the local water utilities be kept as low as possible. Therefore, the Pesticide Use Data Base information was used to determine which substances should be analyzed by each water supply system, substantially reducing the number of analyses required. The lowered cost would not have been possible without the Pesticide Use Data Base. The study probably would have been even more cost effective had hydrogeologic information been available.

The Environmental Protection Agency is in the final stages of developing a monitoring program for pesticides. This is a joint effort within EPA between the Office of Pesticides and the Office of Drinking Water under both the FIFRA and the SDWA. Initiation of the collection of specimens is scheduled for June 1986. In its monitoring program, EPA used data on environmental fate to select the list of chemicals for which they will sample. Included in their criteria is (1) a solubility greater than 30 ppm; (2) a soil sorption coefficient less than 5 mL/g; or (3) previous detection in ground water. Pesticides with usage of less than 1 million pounds per year nationwide were not considered.

On the basis of these data, EPA selected 90 pesticides to monitor. These pesticides are listed in Table 4.5. State programs should certainly consider all of the pesticides listed in Table 4.5, but they should not overlook other pesticides, especially those that were excluded by the EPA because of low-volume usage nationally. The EPA has developed new analytical procedures so that only five chemical analyses were required to detect the 90 chemicals (H. J. Brass, Office of Drinking Water, memo to S. Cohen, Office of Pesticide Program, EPA, September 6, 1985).

The New York State Department of Environmental Conservation has recommended that pesticide companies bear the cost of monitoring for pesticides that have the potential to pollute ground water. Shifting monitoring costs to the public sector encourages diseconomies since neither the farmer nor the pesticide manufacturer are paying the full economic cost for using leachable pesticides. The California bill AB 2021 originally required pesticide manufacturers to pay for monitoring. Due to pressure from the pesticide manufacturers, the bill was amended so the monitoring cost is borne by the state rather than by the pesticide manufacturers. However, AB 2021 is a very significant change in the approach taken by states in monitoring pesticides in that (1) it requires routine monitoring for pesticides identified as potential ground water contaminants; (2) it requires mandatory cancellation if monitoring results indicate that any agricultural use causes significant contamination; and (3) the burden of proof is on the manufacturer to show that the contamination level detected is not of concern.

Reduction in Pesticide Use

Pesticide use can be reduced by (1) implementation of alternate crop production patterns and techniques that require less pesticide; (2) modifications in agricultural practice, application equipment, and timing of changes in the formulation and use pattern of a pesticide product to reduce the likelihood it will leach; (3) replacement of leachable pesticides with materials that are less mobile, persistent, and toxic.

Integrated pest management (IPM) is a term that describes the use of a variety of pest control techniques either singly or in combination to develop more cost-effective management of pest populations. Because IPM promotes the use of nonchemical means of pest control and improves the efficacy of pesticides by improving the timing and placement of applications, IPM can usually reduce pesticide use as well as reduce production costs.

TABLE 4.5 Pesticides Recommended to be Included in EPA National Survey of Ground Water—Chemicals

1. Acifluorfen
2. Alachlor
3. Aldicarb
4. Aldicarb sulfone
5. Aldicarb sulfoxide
6. Ametryn
7. Ammonium sulfamate
8. Atrazine
9. Atrazine, hydroxy-
10. Atrazine, desalkyl
11. Baygon (propoxur)
12. Bentazon
13. Bentazon, 6-OH
14. Bentazon, 8-OH
15. Bromacil
16. Butylate
17. Carbaryl
18. Carbofuran
19. Carbofuran, 3-OH
20. Carbofuran, 3-keto
21. Carboxin
22. Carboxin sulfoxide
23. Chloramben
24. Chlordane
25. Chlorothalonil
26. Chlorothalonil, 2- or 4-OH
27. Cyanazine
28. Cyanazine, methylpropionamide (H_2O addition product)
29. Cycloate
30. Dalapon
31. DBCP
32. Dacthal/DCPA
33. Diazinon
34. Diazoxon
35. Dicamba
36. Dicamba, 2-OH
37. 2,4-D
38. 1,2-dichloropropane
39. Dieldrin
40. Dimethipin (Harvade)
41. Dinoseb
42. Diphenamid
43. Disulfoton
44. Disulfoton sulfoxide
45. Disulfoton sulfone
46. Diuron
47. EDB
48. Fenamiphos (Nemacur)
49. Fenamiphos sulfoxide
50. Fenamiphos sulfone
51. Fluometuron
52. Fonofos (Dyfonate)
53. Hexazinone
54. Maleic hydrazide
55. MCPA
56. MCPA-creson (4-chloro-o-cresol)
57. Methomyl
58. Methomyl metabolites
59. Methyl parathion
60. Methyl paroxon
61. Metolachlor
62. Metribuzin
63. Metribuzin DA (deaminated)
64. Metribuzin DK (diketo)
65. Metribuzin DAKA (deaminated diketo)
66. Nabam
67. Nabam metabolite-ETU
68. Nitrates
69. Oxamyl
70. Oxamyl metabolite-DMCF (N,N-dimethyl-l-cyanoformamid)
71. Paraquat
72. PCNB
73. PCP
74. Picloram
75. Prometone
76. Prometone desalkyl
77. Pronamide
78. Pronamide metabolite–the methylketone
79. Pronamide the xazole metabolite
80. Pronamide the xazole metabolite
81. Propham
82. Simazine
83. Simazine-2-hydroxy
84. Simazine-desalkyl products
85. Treflan
86. Triallate
87. 2,4,5-T
88. 2,4,5-TP
89. Tebuthiuron
90. Terbacil

SOURCE: Cohen, 1985.

Nonchemical means of pest control include (1) the use of plant varieties that are resistant to pest damage, (2) cultural methods such as physical destruction of crop residues (Reynolds et al., 1975) or early harvesting (Shoemaker and Onstand, 1983), and (3) biological control by predators, pathogens, and other natural or introduced enemies of the pest. Another means of nonchemical pest control is "microbial pesticides" like *Bacillus thurengiensis* (BT). BT, a pathogen that attacks lepidopterous insects, is applied like a chemical pesticide. A classic example of a successful integrated program has been developed in Texas for control of insect pests of cotton. The program is based on the use of a new variety of cotton that develops more quickly and hence can be harvested before the boll weevil can cause serious damage. The program also promotes the use of a type of pesticide and a timing for applications that are designed to protect as much as possible the natural enemies of the boll weevil. Masud and Lacewell (1985) reviewed cotton insect IPM programs throughout the United States and concluded that IPM is profitable and reduces pesticide use. In a review of six on-farm insect, disease, and weed IPM programs, Lacewell and Masud (1985) concluded that all reduced insect treatments. Profit was increased in all cases, the increases ranging from $3 to $186 per acre. They also state that IPM control strategies for major insect pests in Michigan increased growers' profits between 16 and 46 percent.

Integrated pest management programs have not been developed for all pests. Although a great deal of work has been done for foliar insect pests, research on IPM for nematodes and weeds has only begun recently and relatively few programs have been developed and implemented. From the point of view of ground water contamination, this is unfortunate since pesticides associated with nematode and weed control tend to be among the most commonly found pesticides in ground water. Solubility and persistence, and the chemical characteristics that promote effective control of nematodes and weeds, are also characteristics that increase the potential of a pesticide to reach ground water. Crop rotation is one method to reduce the need for nematicides.

Integrated Pest Management (IPM) programs for nematodes and weeds in cotton and alfalfa have recently been developed, and they do show promise for cost-effective pest control with a reduced level of pesticide use. Given the demonstrated success of IPM in reducing pesticide use for foliar insect control, research programs should be supported to develop IPM methods for other pests that are responsible for the application of pesticides that have the propensity to migrate into ground water. Such programs usually take several years to develop and many more years to be widely implemented. Hence, although IPM is a very attractive long-term alternative, shorter-term methods for pesticide use reduction must also be considered.

California has attempted to encourage the use of existing IPM methods by requiring all applicators of restricted-use pesticides to provide sufficient information to indicate whether they were using available IPM methods. Such a system depends upon the commitment of the local enforcing agency to maintain close scrutiny of pesticide use to ensure that available IPM methods are implemented.

In some cases, it is possible to reduce leaching of pesticides by changing the timing of irrigation and pesticide applications. Models of pesticide movement indicate that the time interval between the application of a pesticide and the first infiltration of water (either from rainfall or from irrigation) can have a tremendous impact on the pesticide concentration reaching ground water. This is because many pesticides will degrade when they are above the soil or near the surface. The degradation rate generally decreases as infiltrating water moves the pesticide deeper in the ground. Hence, pesticide pollution could potentially be reduced by improved timing of irrigation and by avoiding pesticide applications when rain is predicted. The use of "chemigation," the addition of pesticides to irrigation water, should be considered very carefully. The California bill AB 2021 has a provision for cancellation of registration for lack of data on pesticides that are applied by chemigation or ground injection.

Another way to reduce pesticide pollution in ground water is to replace currently used pesticides with substitutes that are either less toxic or are less likely to migrate into ground water. New pesticides research has resulted in the creation of some pesticides that are less persistent and/or less toxic to humans. Other pesticides are being designed to be applied at very low dosage. However, the high cost for development and toxicological testing of a new pesticide sometimes discourages pesticide manufacturers from marketing a new product that will compete with their own existing pesticide that is selling well. If states or the federal government were to put a tax on leachable pesticides to pay the cost for monitoring (in vulnerable areas) for all pesticides with the propensity to leach, there would be an economic incentive for the pesticide manufacturers to develop and market new herbicides and nematicides that would have different chemical characteristics and would not pose potential problems of ground water contamination.

Another method to reduce the amount of pesticides reaching ground water is to put restrictions on the use of leachable pesticides near drinking water wells and during rainy seasons. For example, Florida has prohibited the use of aldicarb within 600 feet of wells used for drinking water and also prohibited its use before April. Such methods will not eliminate the presence of a pesticide in drinking water, but they can reduce the concentrations to a point that is considered acceptable.

Ground water contamination problems may be exacerbated in some regions by the use of minimum tillage cropping practices designed to reduce soil erosion. Weeds left after minimum tillage are killed by herbicides rather than by plowing as in a conventional tillage system. Minimum tillage results in reduced erosion and surface runoff, which can lead to increased leaching. A possible result would be increased amounts of soluble pesticides reaching ground water. The impact of minimum tillage on pesticide leaching is a good research topic; conclusive data on this subject are not currently available. Minimum tillage is an example of a production technique that reduces one type of pollution (sediment) but may increase another type of pollution (pesticide residues reaching ground water). There are other erosion control techniques, such as contour plowing, that do not cause increased pesticide use. The costs and benefits of alternative erosion control methods vary with crop, slope, soil type, and other factors (Hinkle, 1985).

Other Nonpoint Sources

In addition to pesticides, nonpoint sources of ground water contamination include nitrates (from agriculture, lawn care, or septic tanks), other components of septic tank seepage, improper hazardous waste disposal by homeowners and small firms, mining wastes, road salts and oils, urban surface runoff pollutants, and seepage of polluted surface water. Nitrate contamination of ground water from septic tank seepage, fertilizer applications, and animal feed lots has been well documented.

A more recent problem has been aquifer contamination with organic solvents such as trichloroethylene. Although some nonpoint source contamination from trichloroethylene is caused by improper waste disposal by diffuse small businesses (like dry cleaners), it may also be caused by the use of septic tank cleaners. Septic tank cleaners containing trichloroethylene have been banned in Long Island and Connecticut.

Besides banning materials from use, the primary techniques available to reduce nonpoint source pollution from nitrates, solvents, and industrial chemicals include land use control, substitution of alternate chemical formulations, legislation to control land application of materials, collection of small amounts of diffusely distributed wastes, and public education about proper disposal procedures. A data base describing the distribution of the storage, use, and disposal techniques for small businesses like dry cleaners would also be helpful. Unlike agricultural pesticides, the land application of fertilizers, road salts, and other potential ground water contaminants is not regulated by law. Public education efforts aimed at informing individuals and small businesses of the dangers associated with improper hazardous and other waste disposal as well as informing them of the correct procedures for disposal can be effective in reducing local ground water contamination.

It has been estimated that approximately 30 percent of the population of the United States use septic systems. The volume of waste water processed by septic systems is about 3.5 billion gallons per day. Since the contaminant load from septic systems is proportional to the population serviced by septic systems, restricting septic systems to low-density areas can reduce ground water pollution. Townships on Long Island have implemented strict zoning requirements that prohibit the use of septic systems in areas with densities above a location-specific maximum density. The maximum density ranges from one half acre per dwelling unit in some areas to 5 acres per dwelling unit in the Southampton portion of the Pine Barrens. The town of Brookhaven implemented 1-acre and 2-acre zoning for septic systems on property in the Pine Barrens that previously had been allowed septic systems on smaller lots. The 2-acre upzoning has been upheld following challenges by landowners in both federal and state courts.

The Suffolk County Planning Department, the Long Island Regional Planning Commission, and the townships, in general, recognize that limiting development to protect ground water through zoning and other forms of development restrictions is a high priority. Long Island transferred 644 acres of oak brush plains to the New York State Department of Environmental Conservation (DEC) in August 1984 with a commitment from the DEC commissioner that it be established as a nature preserve.

It is also possible to reduce nonpoint source pollution of ground water by altering the form in which contaminating chemicals are used. The best example is the replacement of ordinary fertilizers with slow-release formulations. Plants cannot absorb all the nutrients in the fertilizer quickly. A slow-release formulation, which is geared to a plant's uptake rate, can reduce the amount of fertilizer applied and the amount of nitrate that leaches. Nitrate leaching can also be reduced by splitting the amount of a regular-formulation fertilizer between two applications. Crop rotation with nitrogen-producing legumes can reduce the need for commercial fertilizers and subsequent leaching. Unfortunately for some situations, crop rotation is not as profitable as producing a monoculture.

Several states have established programs to collect small amounts of hazardous materials from home owners and businesses. For example, Florida has "Amnesty Days," when individuals or small firms can bring solvents, pesticides, or other organic chemicals to central collection areas, where they are disposed of by public authorities. Similar programs have been implemented in New York and California.

Conclusions and Recommendations

Nonpoint sources of ground water contamination are very difficult to control for several reasons. The committee chose to concentrate its recommendations on the use of pesticides because the task of preventing pesticide contamination is very difficult and because of the large number of pesticides in use, the wide range of chemicals, the toxicological and environmental-fate characteristics displayed by these materials, and the lack of information on their environmental fate and health effects. Therefore, the committee recommends the following:

- ***Registration of Critical Pesticides*** **States should consider initiating a routine procedure for flagging pesticides that have potential for leaching into and contaminating ground water. Such a procedure should be based on the pesticide's chemical characteristics and other factors such as evidence of previous detection in ground water. States should consider canceling the registration of pesticides for which essential data have not been provided.**
- ***Data Base on Agricultural Chemical Use*** **States should maintain a data base on the spatial and temporal distribution of applied pesticides. Applicators could be required to report when, where, and how much of these pesticides is applied. Also useful are maps and summaries to indicate where such materials are applied.**
- ***Pesticide Tax to Fund Monitoring*** **Monitoring should be used to ensure that currently registered, potentially leachable pesticides do not reach ground water. States should consider funding through fees paid for pesticides or their use. Such a program has two advantages: (1) a cost more reflective of the true cost of the pesticide is then paid by its users; and (2) there is economic incentive for the manufacturers to produce new pesticides that do not have the potential to leach into ground water.** An alternative to direct charges on manufacturers is a tax on the pesticide paid by the user. Such a tax would not only fund the necessary monitoring, but could encourage consumers to replace the leachable pesticide with a nonleachable substitute.
- ***Cancellation of Pesticide Registration in Local Areas*** **States that are reluctant to cancel the statewide registration for a potentially leachable pesticide should consider canceling registration in local areas where soil conditions or other factors indicate that pesticide leaching may be a serious problem.**
- ***Economic Incentives, Legislation, and Financial Support for Source Reduction*** **States should encourage the use and development of source reduction techniques such as pesticide substitution, changes in irrigation practices, prevention of pesticide and fertilizer application near drinking water wells, and integrated pest management.**

Source Reduction

One of the most effective long-term strategies for ground water protection is to reduce and/or eliminate the sources of contamination. With today's technology and knowledge, it should be possible to substantially reduce the quantity of contaminants discharged to the ground. Current government policies regarding environmental management, the increasing cost of waste disposal, increasing legal and financial liability for damage to the environment from current handling, storage, transportation, and disposal of hazardous materials and waste, and changing societal attitudes toward chemical usage and waste disposal provide strong incentives to reduce contamination sources.

There are a number of source reduction strategies that have been and can be employed in the future to protect ground water. Following is a summary of these strategies with examples from some of the state and local ground water protection programs examined by the committee. This summary is neither comprehensive nor exhaustive. The reader is advised to consider a growing body of literature on this subject such as the National Research Council (1985) report entitled *Reducing Hazardous Waste Generation*.

Prohibition of Activities

Prohibition of polluting activities is an effective means of source reduction. This includes elimination of ground and underground discharges, banning usage of potentially polluting products, and prohibiting certain activities in important ground water recharge areas.

For example, the Suffolk County Health Department has banned the use of organic septic system cleaners throughout the county. Land disposal of solid and hazardous waste is also prohibited in deep water recharge areas on Long Island. The New Jersey Pinelands Commission has prohibited the storage of certain hazardous materials and the land disposal of hazardous and solid waste in the Pinelands area of southern New Jersey. The Florida Department of Environmental Regulation prohibits the land disposal of hazardous waste and any new deep-well injection of hazardous waste in Florida. California has banned a number of broad-use pesticides such as ethylene dibromide (EDB) because of widespread ground water contamination from these pesticides.

Changes in Industrial Operations

Industrial processes that eliminate or reduce the quantity of waste produced can be substituted for older, more-polluting industrial processes. For example, replacement of solvent-based paints by water-based paints in the automotive industry has significantly reduced the amount of waste solvents and sludges to be disposed. Reduction in the toxicity of waste can be accomplished through simple in-plant treatment. Reduction of the quantity of waste can be accomplished through good housekeeping practices or by technology to concentrate or treat waste resulting in a significant reduction of waste streams that would otherwise find their way onto the land or into ground water. For example, caustic and acidic waste streams can often be neutralized to eliminate their polluting potential. Separation of waste streams often allows recovery of products from such waste streams.

Industrial and Consumer Product Substitution

Substitution of less-polluting industrial and consumer products can significantly reduce the potential for ground water contamination. For example, the Cape Cod Planning and Economic Development Commission has sponsored educational programs for small industries and individuals on strategies for reduction of hazardous waste. There are approximately 200 consumer products used in the home that contain toxic or hazardous substances. Less-polluting substances can be substituted for many of these products. For example, latex (water-based) paint can be used instead of oil-based paints; vinegar and water can be used to remove some stains from clothing instead of solvent-based stain removers; detergent-based cleaners can be used instead of solvent-based household cleaners.

Recycling and Reuse of Chemicals, Petroleum Products, and Waste

Recycling and reuse of chemicals, petroleum products, and waste have been successful in limiting ground water contamination. With little or no modification, waste can often be reused by the original industrial facility or by other industries. For example, waste solvents from the electronics industry have been used in the manufacture of paints. Waste streams often have economic and other value if they are recovered and treated through processes such as distillation. For example, waste oil, which has often been disposed of through discharge into sewage systems and into landfills and used as a dust suppressant, can be collected, refined, and reused. Used solvents can be distilled and refined for reuse rather then disposal. Recycling and recovery of glass and metals before disposal of solid waste can reduce the quantity of waste to be disposed of and eliminate ground water contamination. Further recovery of these sources in solid waste can be accomplished by converting waste to energy through thermal decomposition. These methods of resource recovery can eliminate up to 80 percent of the solid waste normally disposed of on land. Caution should be exercised in employing trash-to-energy resource recovery, however. Air emissions of heavy metals, toxic organics, and acid gases can pose a health threat if they are not well controlled. As stated earlier in this chapter, Connecticut, Cape Cod, Massachusetts, New Jersey, and New York have active programs to promote trash-to-energy resource recovery facilities for solid waste disposal.

Alternative Waste Disposal Strategies

Employment of alternative waste disposal strategies such as neutralization, treatment before discharge to ground, and thermal decomposition can eliminate ground water contamination. Many large industrial facilities and chemical plants are moving rapidly to eliminate deep-well injection by using alternative waste disposal strategies. Some of the programs examined by the committee have also encouraged alternative waste disposal strategies for small-quantity hazardous waste generators and household hazardous waste. The Cape Cod Planning and Economic Development Commission (CCPEDC) and the Suffolk County Health Department have established household hazardous waste collection and disposal programs that helped eliminate disposal of such waste at local landfills. The CCPEDC has recently initiated a technical assistance program for management of hazardous waste from small-quantity business and industrial generators for waste reduction and disposal at environmentally sound disposal facilities. Florida has conducted several "Amnesty Days" for collection of individual and institutional hazardous waste that would have probably been disposed of in a local landfill.

Factors Affecting Source Reduction

A number of factors can affect society's ability to reduce sources of ground water and other environmental contamination:

- Access to information on strategies for reducing waste and other polluting activities.
- Access to funds for capital investment in new equipment or programs for waste reduction.
- Social, political, and economic goals that determine the actions of institutions and individuals.
- The cost and availability of alternative strategies for waste disposal.
- Government regulations and incentives for waste reduction and elimination.
- Financial liability for damage to individuals and the environment and/or for remedial activities because of improper waste disposal.
- Public attitudes and opinions.
- Availability of technology to reduce or eliminate waste and other polluting activities.

Conclusions and Recommendations

It is obvious that a long-term strategy for ground water protection is to reduce and/or eliminate the sources of contamination. There are cases where source reduction offers a potentially powerful and cost-effective means of minimizing ground water pollution. In general, the state programs examined are weak in source reduction programs such as waste incineration, recycling, and better management practices; it is evident that additional incentives are needed to accelerate and expand source reduction efforts by industry and the public. Therefore, the committee recommends the following:

- **States should consider regulatory and economic incentives for source reduction by industry, government, commercial interests, and the public. States should also consider a variety of financial assistance programs to encourage waste reduction in industry, such as low-interest or no-interest government loans for capital cost of new equipment or environmental audits to determine the best way of reducing waste generation; tax reductions or credits; grants or other aid to encourage smaller firms to pool resources and implement a joint waste reduction strategy; government subsidies to firms actively working on new methods of reducing waste; financial assistance to waste exchanges to encourage more recycling and reuse of materials that might otherwise be disposed of on the land or into ground water.**
- **State agencies, university-based groups, trade associations, and other institutions should develop educational programs for local industries and the public to disseminate information on waste reduction technology and assist them in implementing waste reduction practices. Specific emphasis should be on medium- and small-sized generators of industrial waste, which do not have the expertise or time to keep abreast of technological innovation.**
- **The committee believes that EPA should fund additional research on source reduction technologies. EPA should also fund programs that include research into public and private practices in the use of substances that are potential ground water contaminants.**

Prohibition of polluting activities is one of the most effective means of source reduction. This includes eliminating ground and underground discharges, banning the use of potentially polluting products, and prohibiting certain activities in important ground water recharge areas. Several areas of the country have used prohibition of polluting activities successfully through source regulation or land use controls. For example, Suffolk County (New York) and the state of Connecticut have banned the use of organic septic system cleaners.

The committee found that several state and local entities use hazardous and solid waste disposal strategies other than land disposal. For example, programs such as Florida's "Amnesty Days" have successfully collected household hazardous waste and hazardous waste from small-quantity industrial and commercial generators that otherwise might have been improperly disposed of. However, few states or communities have adopted programs for aggressively promoting source reduction of hazardous waste and recycling of solid waste.

- **The committee, therefore, recommends that all state and local entities consider similar strategies for reducing improper disposal of household and other small-quantity generator hazardous waste. Municipalities and states should also consider the relative merits of comprehensive solid waste recycling and incineration programs.**

Land Use Controls

Almost every activity on the land surface has some potential for contaminating the underlying ground water whether it is nonpoint sources such as residential lawn fertilization, the use of a septic tank, or highway runoff or point sources such as a discharge from an industrial lagoon or a hazardous waste disposal site. The degree of risk to ground water quality is controlled by two basic factors:

1. The susceptibility or sensitivity of the aquifer to contamination, such as a very shallow water table overlain by permeable sands.
2. The potential types, magnitudes, and locations of contaminant discharges to the aquifer from the specific land use activity. For example, an industry that uses large amounts of chlorinated organic solvents could potentially have a greater adverse impact on ground water than residential septic tanks or lawn fertilization.

Contamination sources located within the area of water recharge contribution of a wellfield in a shallow water table aquifer pose maximum risk. Lower risks are posed if the source is a substantial distance upgradient or is downgradient from the well. A source that is close to a well can be within the same aquifer, posing a high risk, or can be separated from the well by a geologic confining unit or a flow system boundary. An understanding of these relationships forms the principal basis for the land use approach to ground water protection. Certain land uses or activities can be restricted or prohibited within designated critical areas. All of the above factors should be considered in the delineation of areas where special controls will be applied.

The land use approach is closely linked to the concept of ground water classification; that is, different aquifers or portions of aquifers require different levels of protection. Land use controls are typically applied on a local level and frequently involve more detailed mapping and zoning of land areas or aquifers than state classification systems employ. Moreover, land use controls may take additional factors into consideration, such as economic impacts of protection measures and consistency with existing zoning and current development patterns. State classification and mapping programs may provide the overall framework for land use controls, but the final designation of critical areas for local land use regulation may differ from the state-level efforts.

Land use control programs are generally most effective at protecting shallow unconfined aquifers. Many of these measures such as wellfield restrictive zones would have little or no application to protecting deep confined aquifers. However, one land use control that can be effective for confined aquifers is the establishment of protective zones on surficial recharge areas, such as has been done on Long Island.

Land use controls are implemented through the application of various forms of local and regional authority to protect the public health, safety, and welfare. These include regulations and ordinances adopted by the local health agency, zoning by-laws, and municipal ordinances enacted under general home rule authority. In highly critical areas, preservation of hydrologic conditions in their natural state is required. Open land preservation can be accomplished through public and private purchase, eminent domain in tax lien takings, and conservation easements.

Successful land use management in critical areas requires ongoing planning activity. Site plan reviews during major land developments and property conversions can accomplish many of the objectives of ground water quality protection. Sanitary and industrial surveys and mapping of aquifers, water supplies, and contamination sources are important tools in establishing appropriate land use controls. As new information becomes available, boundaries of critical areas can be revised. Monitoring data should be used to identify problems and assess the effectiveness of controls.

In many states, local authority has been more effective in controlling ground water pollution problems than state authority. For example, county and municipal ordinances controlling underground storage of gasoline were in place in Cape Cod, Massachusetts, Long Island, New York, Dade County, Florida, and Santa Clara County, California, one to five years before state or federal laws were enacted.

Local land use controls to protect ground water quality have been implemented in numerous communities in Massachusetts, New Jersey, Connecticut, and Florida. Examples of unique or innovative local programs were found in Cape Cod, Long Island, and Dade County. This section will focus on several components of these programs: designation of critical resource areas, zoning by-laws, health regulations, county and municipal ordinances, and state-level activities.

Critical Area Delineation

At the heart of the land use approach to ground water protection is the delineation of critical areas. Various methods and criteria have been employed, ranging from designation of entire geologic units to the delineation of areas of contribution for individual wells in local zoning programs. The complexity of the hydrogeologic setting and extent of the data base are major factors determining the method of critical area delineation. The aquifers in all three of the counties examined (Long Island, Dade County, and Cape Cod) were designated as sole-source aquifers under the federal Safe Drinking Water Act.

Hydrogeologic Zoning on Long Island The Long Island Regional Planning Board (LIRPB) created the concept of hydrogeologic zoning as the basis for local critical water protection area delineation. The hydrogeologic setting is relatively complex, with several interconnected aquifers. The data base is also extensive, built on more than 30 years of study by USGS and others. Through an understanding of the hydrology, areas were identified that contribute recharge water to the deep aquifer system. The hydrogeologic zoning scheme classifies those deep recharge areas as zones requiring critical protection. In addition to recharge conditions, current land use, ground water quality, and surface and ground water relationships were used to subdivide the recharge zones.

The LIRPB works with county health departments and the state DEC under planning grants to define the critical hydrogeologic zones in greater detail. In addition, USGS studies have provided information supporting the modification of the conceptual zones themselves. These developments illustrate several important factors in designation of critical areas. Ongoing planning and data collection have been necessary to refine the critical area designations. The fact that the boundaries require refinement over time has not weakened the usefulness or public acceptability of the concept on Long Island.

Not only have the hydrogeologic zones been updated, they have been subdivided and refined for different purposes. Working in cooperation with the New York DEC, the LIRPB has identified Special Protection Areas within the deep flow recharge zones. These are the portions of the deep flow zones in which the land is relatively open and preventive planning will have the greatest impact.

Another critical water protection area, the Pine Barrens area of Long Island, has been designated by special county legislation. This 100,000-acre area, slated for preservation of both water quality and a unique ecosystem, was delineated on the basis of vegetation. It roughly coincides with the deep flow recharge zones.

The Suffolk County Department of Health Services has defined additional Water Supply Sensitive Areas including zones around public supply wells screened in the upper glacial aquifer. These zones are based on the water budget area for the supply well and extend 1500 feet upgradient and 500 feet downgradient of the well. They also include areas where the upper aquifer is underlain by salt water. These areas are mapped by the Suffolk County Health Department.

Wellfield Areas of Contribution—Cape Cod, Massachusetts The Cape Cod Planning and Economic Development Commission (CCPEDC) has worked with the 11 local communities on Cape Cod that have public water systems to delineate zones of contribution for their wellfields. In the Cape Cod case, vertical flow is not as important an issue as it is on Long Island. USGS studies and modeling efforts have established that although the glacial aquifer is up to 350 feet thick, the predominant flow is through the upper 100 feet, which is the zone tapped by public water supply wells. Assuming a relatively uniform flow system, the planning agency has used basic hydrogeologic principles for calculating pumping drawdowns to delineate Zones of Contribution (ZOC) around wellfields. The boundaries based on estimated hydrogeologic parameters are admittedly approximate but have been accepted by communities as the basis for zoning ordinances. As with any planning program, the boundaries are periodically updated as new data become available.

Wellfield Protection Ordinance: Dade County, Florida A unique approach to critical area delineation has been developed by the Metropolitan Dade County Department of Environmental Resources Management and the Dade County Planning Department. Funded through a Section 208 planning grant to the three counties dependent on the Biscayne aquifer for water supply, the project involved the use of a mathematical ground water flow model as a tool for predicting the recharge area of influence for individual wellfields. The model was also used to provide information on ground water travel time to the wells under pumping conditions.

The Florida Biscayne aquifer, underlying the greater part of Dade, Palm

Beach, and Broward counties, is a highly prolific water table aquifer in porous limestone. No more than a few hundred feet thick, the aquifer is capable of yielding thousands of gallons per minute to individual wells. Over 300 million gallons per day are withdrawn for public water supply. The aquifer is highly susceptible to contamination from the surface and to salt water intrusion from overpumping near the coast. In 1977 and 1978 the majority of public supply wells were found to have detectable levels of synthetic organic chemicals.

Within the area of influence around the large wells tapping the Biscayne aquifer, ground water flow velocities range from 1 to 3 feet per day in the outer boundaries of the area and increase to 100 feet per day in the steep gradient area near the wells. The Dade County study concluded that more stringent controls were necessary within the inner zones where contaminants would move rapidly toward the wells.

A multitiered protection system was developed, ranging from total protection within 100 feet of the well to lesser levels of protection between 100 feet and the 10- to 30-day travel time distance and 30- to 210-day travel time distance. The inner zones are based on the die-off of bacteria in soil and the ground water environment. A second consideration was dilution provided by recharge. The 210-day travel time interval is the longest period with no rainfall on record. Thus, protection of the 210-day zone was viewed as protecting all the ground water that could reach the well undiluted by recharge. The 210-day travel time radius is up to 2 miles for a large wellfield.

Key provisions of the original ordinance include the following:

- Prohibition on discharges within a 10-day travel time zone.
- Establishment of different zones based on travel time to public supply well "cones of influence" defined as the 210-day travel time.
- Limitations on sewage loading to septic tanks based on presence of sandy substrata together with travel time to well (ranging from 140 gallons per capita (gpc)/acre within 100 feet of well to 850 gpc/acre within a 210-day zone of influence where indigenous sandy strata are not present).
- Limitations on waste water effluent for residential and nonresidential property served by sanitary sewers (maximum of 1600 gpc/acre within 30-day travel time; 850 gpc/acre between 100 feet and 10-day travel time.
- Prohibition of hazardous materials use, generation, handling, and storage within the zone of influence of wellfields. The prohibition of hazardous materials is enforced through a covenant to be given in favor of the county placing this restriction on a parcel of land. The covenant is required before issuance of a building permit or occupancy permit.

The travel time concept proved to be controversial for several reasons (Yoder et al., 1984). Conservative (nonreactive) chemicals such as nitrate from sewage disposal are not removed or delayed by travel through the aquifer. Moreover, the major class of contaminants of concern in the county, synthetic volatile organic compounds, undergoes very little degradation in the ground water system. In view of this information, the travel time was viewed as arbitrary. The initial zones were also found to be too small, as they did not protect all the recharge area within the zone of influence of the wellfields.

Long-range plans for Dade County's water supply call for the eventual abandonment of most existing municipal wellfields and use of only three major county wellfields, one of which is new. Since 1983 the Department of Environmental Resources Management (DERM) has conducted a comprehensive study to formulate a management plan for the three major wellfields that will supply the county with its future water supply. The county felt that the original ordinance was not adequate to protect the pristine character of one of the relatively undeveloped wellfields, particularly in view of the pressures for industrial development in the area. The widespread nature of organic chemical contamination also contributed to the concern that the original ordinance was not adequate.

To investigate this concern, the county formed a special study group that evaluated causes and influence of contamination in the wells and found that very low or moderate levels (range of 0.1 to 4.6 ppb total) of volatile organics were detected when ground water flowed through residential, commercial, institutional, recreational, and utility and transportation land uses and inland water. Ground water flowing through industrial areas had higher levels than in any other type of land use (13.1 to 57.3 ppb). These data were useful in gaining support for a revised version of the ordinance that would place prohibitions on industry throughout substantially larger cones of influence than mapped in the original program. The study also evaluated the cost and availability of treatment technologies for contaminated drinking water supplies as an alternative to prevention. Finally, alternative methods of delineating cones of depression based on different surface water and ground water use scenarios were evaluated.

Once the revised wellfield protection areas were identified, it became apparent that the prohibition on all storage, handling, generation, and discharge of hazardous materials would affect very large areas. It would essentially eliminate industry throughout the 82.5-square-mile protection area designated for one of the wellfields. In addition, the prohibition did not address the existing industrial uses within two of the wellfields. The ordinance was therefore revised to allow certain uses that employ small volumes of materials under carefully controlled conditions. Additional amendments were made to require connection to public sewers and for setting minimum lot sizes for septic tanks. The key provisions of the revised code are as follows:

• Prohibition of hazardous materials applies throughout the maximum day pumpage protection area for one wellfield and the average day pumpage area for the others.

• Connection of public sewers is required if the nearest connection is within 100 feet of the property or meets a formula based on square footage of the building.

• An operating permit is required for all land uses within the maximum protection areas that will discharge treated waste water by any method other than public sewers or that use, generate, handle, dispose of, discharge, or store hazardous materials on any portion of the property. This covers uses such as electrical transformers, cleaning vehicles, or outdoor storage that would not require a building permit.

• Small generators of hazardous waste are allowed provided that the following prevention and monitoring measures are taken:

— Monitoring.

— Secondary containment, inventory control, and record keeping of hazardous materials.

— Storm water management of water pollution caused by hazardous materials.

— Protection and security of facilities utilized for storage and handling of hazardous materials.

• A 50 percent increase in the volume of hazardous materials used at existing sites within the protection areas is allowed provided that the facility is upgraded to reduce the risk to the environment.

• Minimum lot sizes for single residences on septic tanks are established.

In summary, the revised ordinance represents a compromise between economic growth and ground water protection. Placing small generators under strictly controlled conditions allowed the county to place a significantly larger area under protection. Many of the revisions contained in the new ordinance could only have been developed through experience in administering the initial version.

Ground Water Development Restrictions A corollary approach to wellfield protection measures, such as those of Dade County, Florida, and Long Island, New York, would be restrictions on placement of wells near known contamination zones. If a wellfield is allowed to be placed near an existing zone of ground water contamination, the well(s) can draw contaminants into their previously clean zone of production, thus expanding the area of contamination. Although none of the programs examined by the committee had such restrictions explicitly in place, they could easily be made a part of well or withdrawal permitting programs.

Zoning By-Laws

In each of the local programs reviewed, some form of zoning has been used to accomplish basic ground water protection objectives. Land use zoning employs local police powers to control a variety of activities under authority granted through state enabling legislation. Its primary impact is in areas that are undeveloped at the time of the rezoning; however, certain performance controls over specific activities such as waste water discharge or handling of toxic and hazardous materials can at present have an impact on newly developed land at the time of development and use.

Zoning for Increased Lot Sizes Lot size increases and housing density restrictions can be incorporated directly into existing zoning districts if the districts fall within a critical water protection area. This approach has been used to change lot sizes in a number of Long Island communities, with protection of ground water quality as one of several objectives.

The Southampton Township has recently amended its Master Plan and Zoning Ordinance to require minimum 5-acre lot sizes on 25,000 acres of Pine Barrens in that township. East Hampton has imposed similar zoning restrictions on approximately 5000 to 6000 acres of critical watershed lands. In addition to upzoning substantial Pine Barrens areas to 1- and 2-acre lots, the town of Brookhaven has also rezoned a large portion of industrial land to residential use within the deep flow recharge zone.

Any move to increase lot sizes substantially tends to raise issues of property confiscation and exclusionary zoning. Several communities on Long Island have faced such challenges. The town of Brookhaven, in a landmark case, successfully defended its 2-acre zoning to protect ground water quality. Because of the highly controversial nature of large-lot-size zoning, several communities on Cape Cod found it practical to separate zoning proposals for lot-size change from proposals for other controls within critical recharge areas.

On Long Island the Special Legislative Commission on Water Supply Needs of Long Island has recommended that towns provide for growth in less-sensitive areas by downzoning (increasing density). The commission has published a map of potential growth centers. In this way a balanced proposal for growth and preservation is presented.

Zoning Overlay Districts Where critical water protection areas do not conform to zoning boundaries, overlay water protection districts are used. These superimpose additional restrictions on existing zones but do not change the underlying zoning. This approach is used for resource preservation in a number of instances, such as flood plain zoning, and is a simpler procedure than changing all the town zoning districts. An advantage of this approach is that the zones can be presented and defended as being based on a single technical factor: ground water protection. This has been considered the strongest defense against charges of exclusionary zoning or confiscation.

One potential difficulty of applying the overlay approach to wellfield protection is that the boundaries tend to be based on some form of modeling or calculations, and the lines would be difficult to locate on the ground. Unlike the Long Island hydrogeologic zones, field measurements could not be used to confirm the boundaries as they are based on assumptions of pumping patterns and recharge conditions that may not exist at a given time. It is up to the zoning enforcement officer to make a determination whether a specific development is within the zone based on a comparison of maps. Overlay district boundaries can also split single parcels or industrial parks, creating inconsistencies in their development and possibly raising equity issues among developers being subject to drastically different requirements within the same geographic area.

For all of the above reasons, some planning agencies, both within Dade County, Florida, and some towns on Cape Cod, have slightly modified the zones of influence on wellfields to take into consideration current land use patterns, property ownership, and existing zoning boundaries. This can be justified to some extent by the additional benefits to the public of special zoning such as public enjoyment of open space and preservation of natural environments. While this action may be viewed as a diversion from the technical basis for the zoning, to date it has not been successfully challenged in these areas. It may be an advantage in that the zones are not adopted strictly on the basis of one piece of information (i.e., pumping drawdown) and therefore cannot be challenged solely on the basis of the technical aspects by which that information was developed.

The Cape Cod zones of contribution have been used as overlay districts encompassing large portions of towns and several underlying zoning districts. A broad range of provisions has been built into zoning overlay districts in the towns under study in Cape Cod and Long Island. These include the following:

- Bans on such activities as landfills, toxic waste disposal, toxic materials storage and handling, and road salt storage.

- Limits on potential nitrate loading. The Cape Cod model by-law sets a performance standard for nitrogen loading, allowing an applicant to calculate nitrogen released from both on-site sewage flows and fertilized turf. The applicant is then provided flexibility to trade off one against the other in development of the site. Design specifications are based on a Southampton, New York, overlay district, which restricts the amount of square footage that can be devoted to fertilized turf and shrubs to no more than 15 percent of the plot. This is based on research by the Long Island Section 208 plan establishing the rates of nitrogen leaching by turf culture.
- Limits on sewage flows, in gallons.
- Natural area preservation. The Southampton, New York, overlay district requires that 80 percent of a lot be left in its naturally vegetated condition to maximize recharge, minimize chemical fertilization, and preserve water quality, and promote water conservation by minimizing watering.
- Runoff management.
- Site plan review. Probably one of the most important requirements of the Cape Cod by-laws, the site plan review allows for a close examination of all aspects of development within the critical zone. The Cape Cod by-law requires a nonresidential applicant to provide information on the type and quantities of chemicals that will be used on the site, measures that will be taken to prevent and contain spillage of the chemicals, and the means and availability of proper disposal of any hazardous wastes generated on the site. Outside experts may be involved in the site plan review to provide expertise not available within a town planning board.
- Special Review Board. The Cape Cod model ordinance creates a special interdepartmental review board with representatives of health, water, and conservation agencies as well as the fire officials who manage permits for underground storage of petroleum products and chemicals. Permits are granted by this special board, providing maximum input of available expertise into the decision.

In summary, zoning has been successfully employed for the protection of thousands of acres to tens of square miles of critical areas on Cape Cod and Long Island and in Dade County, Florida. Through zoning, not only can the density of sources of ground water contamination be reduced but a broad range of controls over future activities can be put into place. The main limitation on the effectiveness of zoning is the degree of development already existing in the recharge area. All three areas (Cape Cod, Dade County, and Long Island) have determined that substantial portions of their critical zones are too developed for zoning to have any impact. Other forms of control, through health regulations and general by-laws, must be invoked to provide the needed protection from ongoing activities in these areas.

Health Regulations Controlling Activities in Critical Areas

Two unique examples of nonzoning approaches to ground water protection by land use controls are provided by Long Island and Cape Cod. Both areas have made extensive use of local health authority to accomplish controls over land use and human activities. These include the following:

- Suffolk County's use of its septic system permitting authority to control housing density and to prohibit certain kinds of industrial development in critical zones.
- Cape Cod towns' use of general by-laws and health regulations to control the storage and handling of toxic and hazardous materials.
- Nassau County's (Long Island) use of its Sanitary Code to control housing density and discharges in special protection areas.

Suffolk County, Long Island The Suffolk County Department of Health Services approves all on-lot sewage disposal systems under standards set by the New York State Sanitary Code and additional county restrictions. Towns cannot issue building permits unless the county's approval has been obtained for the disposal system. The county code prohibits construction of new houses using septic systems on lots smaller than 40,000 square feet within Zones III and VI. In effect, the requirements override local zoning where public sewering is not available.

In 1984, the Nassau County Health Department passed Health Ordinance X, which places similar restrictions on housing density with on-site sewage disposal within Special Protection Districts (see "Critical Area Delineation" earlier in this chapter). In addition, the article bans any disposal of nonresidential waste water on-site within the districts.

Article 7 of the Suffolk County Sanitary Code restricts the siting of industrial operations that use, generate, transport, or dispose of hazardous waste within any of the deep recharge zones (I, II, III, and V). The entire text of the regulation appears as Appendix E to this report. Key features include the following:

- A broadly defined scope of the regulation to include control over nonresidential structures, processes, facilities, and activities.
- A definition of toxic or hazardous chemicals, including a list of nearly 50 specific chemicals and several classes of chemicals. The incorporation of a list of chemicals into this kind of regulation facilitates implementation.

- Prohibition against using or storing toxic or hazardous materials on any site if storage involves quantities greater than 250 gallons or 2000 pounds of dry material. Exceptions are provided for limited storage of fuel products for on-site use. Retail gasoline stations are allowed provided they conform to the strictest provisions of the county code for underground storage tanks.
- Prohibition against discharge of industrial wastes from processes containing restricted toxic or hazardous chemicals to the ground water or to any sewage system. The law contains an exception that allows discharge of such waste to a central sewage system that discharges outside of the deep recharge zones or to the ocean.

It is important to note that existing uses and discharges are not affected by passage of this regulation. Therefore its anticipated effectiveness is largely due to the undeveloped nature of the areas it covers. This is not so great a deficiency in Suffolk County as it might be in other areas because most discharges are already controlled (although not prohibited entirely) by the state ground water discharge permit (SPDES) program. Storage of toxic and hazardous materials is rigorously controlled by the county under its Article 12.

Article 7 will prohibit a large segment of potential industrial development in 260,000 acres, approximately 25 percent of the county. Because of its perceived economic impact, it was strongly opposed by the supervisors of Brookhaven and Riverhead, where much of the undeveloped industrial land is located. Contrary to early fears, the restrictions contained in Article 7 have not driven industry from Suffolk County. Similar restrictions were placed through deed restrictions on an industrial park in the county, which was to be occupied by a major pharmaceutical company. The developer placed the restrictions on the property in response to pressure generated by a major local newspaper. It was found that several businesses were attracted to the location, wanting to be located in a clean site where they would not risk sharing the liability of a polluting neighbor.

In spite of the broad impact the regulation is expected to have on land use, no major new commitment of manpower is planned for its implementation. Article 7 is largely enforced through existing permit programs such as on-site sewage disposal, storage tank regulations, and state discharge permits. Implementation of the by-law has not yet begun, but it appears that this is a resource-efficient means of accomplishing comprehensive protection of ground water from nonresidential uses.

County and Municipal Ordinances

Municipal By-Laws Controlling Toxic and Hazardous Materials on Cape Cod A model by-law drafted by the Cape Cod Planning and Economic

Development Commission under Section 208 continuing planning funds controls the storage and handling of toxic and hazardous materials, without actually prohibiting their use. Key provisions include the following:

- Generic definition of toxic and hazardous materials, such as organic chemicals, pesticides, solvents, paints, and thinners, including waste products containing these materials along with referenced lists published by EPA under the Resource Conservation Recovery Act (RCRA) and the Clean Water Act.
- Required registration with the Board of Health by all nonresidential users of toxic or hazardous materials who store 50 gallons or 25 pounds of such materials.
- Standards setting for storage of materials, including spill containment, similar to Suffolk County's Bulk Storage regulation.
- Prohibition and discharge of any toxic and hazardous materials to the ground.

Finally, it was thought that bringing the article before a town meeting would provide a good opportunity for public education on the importance of ground water protection. The by-law was initially intended for application only in the designated zones of contribution for public supply wellfields. However, Barnstable, the first town to propose the by-laws, rejected this approach. This was because of general discomfort with the means by which the zones were initially drawn and a concern that toxic and hazardous materials should be carefully managed throughout the town. The by-law was passed the second time it was proposed, applicable townwide.

Other towns on Cape Cod have chosen to pass the by-law as a health regulation. This was intended to avoid the uninformed controversy generated by bringing a health-based regulation before a political body (town meeting). In these cases, the by-law was enacted as a health regulation by the appointed board of health, a three- to seven-person body. The wide public acceptance of the by-law or health regulation, which has been passed by 11 of the 15 towns, is attributed to the fact that it does not obstruct or ban industrial development. For many businesses, its only impact was to require information to be filed with the board of health on the nature of activities and materials stored. This information is used to set inspection priorities and is expected to be useful in the future when materials are detected in monitoring programs.

Even in areas where implementation has not been given a priority, the program has contributed significantly to public awareness of the toxic materials problem. Pamphlets prepared under the implementation program emphasized the wide variety of materials and processes that are not generally perceived as heavy industry but pose a threat to ground water quality. Important information was also given to industries on the risks and liabilities associated with their practices and the relatively simple measures that can in many cases be taken to prevent major contamination problems.

Finally, implementation of the by-law or health regulation has resulted in the identification of numerous small generators of hazardous wastes, many of whom had no means available for legal disposal of their wastes. The county is attempting to provide them with information on proper disposal and to put groups of small generators in touch with each other so that they could economically dispose of their wastes with commercial haulers.

County Wellfield Protection Ordinance, Dade County, Florida The Potable Water Supply Protection Ordinance was adopted under the general home rule authority of Metropolitan Dade County in March 1981. Similar to an overlay zoning district, the ordinance imposes standards for land uses within the zones of influence of public supply wells without changing the underlying zoning (Section 24-12.1 of the Code of Metropolitan Dade County, Florida). More detailed explanation of the zone of influence approaches used in the Dade County ordinance has been presented earlier in this chapter (see page 52).

The evolution of the Dade County Wellfield Protection Ordinance illustrates again how ongoing data collection and a renewed understanding of the problem can be used to improve aquifer protection programs. The boundaries have been challenged technically and revised several times, with both technical and policy issues at stake. However, the fundamental concept of the ordinance has not been challenged, and each revision has led to an increased level of protection.

State-Level Programs

New York New York State is in the process of implementing a state-level program to protect "primary" aquifers in all portions of the state other than Long Island. Long Island has its own more restrictive protection programs. The Upstate New York Ground Water Management Plan defines two categories of water supply aquifers: primary and principal. A primary aquifer is one currently being utilized by major municipal water supply systems. Principal aquifers are highly productive formations that are not intensively used as water supplies at present. They are viewed as potential water supplies, but their yield has not been fully established.

The New York Department of Environmental Conservation (DEC) has established as a priority the completion of detailed maps of the primary aquifers on a scale that is usable to all interested parties in the public and private sector. The data base includes county and regional studies published periodically by USGS since the 1930s. A 1980 survey by the Department of Health (DOH) concluded that 7500 people were dependent on ground water provided by public and private ground water supplies in 18 primary aquifers. Implementation of Article 7 has not yet begun, but it appears that this is a resource-efficient means for accomplishing comprehensive protection of ground water from nonresidential uses.

Aquifer maps prepared by state and local authorities in cooperation with the USGS contain detailed information on aquifer thickness, flow directions, well yields, permeability, and land use above the aquifer. DEC has currently received funds from EPA to identify landfills and other threats above the primary aquifers. DEC is also considering requiring the identification of "wellhead areas," or zones of influence around public supply wells, as a condition of public water supply permits for wells supplying 100,000 gallons per day.

Other categories of critical areas called for in the State Management Plan include critical recharge areas and special management areas. Special protection may be required in these areas, owing either to the pristine nature of the water resource or to the stresses to the system due to overpumpage or high level of overlying urban land use.

The New York DEC will rely heavily on mapping, public education, and technical support to local planning efforts to protect the primary recharge areas of upstate New York. The DEC has already presented slide shows and other materials to regional and local agencies in 11 primary aquifer areas. The state is developing a pilot aquifer-specific management plan in one six-county capital district region, modeled after the Long Island plan. The plan will identify threats to ground water quality, necessary land use control and remediation efforts, and the appropriate management agencies. This project will serve as a model for the other primary aquifers.

The DEC is also utilizing its hazardous waste staff to identify facilities in the primary aquifers. In addition, the DEC has identified landfills and active SPDES (ground water discharge permit) holders in these areas. Inspection and enforcement priorities will be set based on this information. Direct state control over land uses in primary ground water recharge areas is called for by the State Watershed Protection Law (Chapter 951 of the laws of 1983). The law requires the commissioner of DEC to promulgate a list of hazardous substances that threaten human health and the environment and to issue rules and regulations that will restrict or prohibit the use or storage of hazardous wastes over primary ground water recharge areas. The law specifies that protection be placed on the deep flow recharge zones of Long Island and may have contributed some momentum to the passage of Suffolk County's Article 7. If funded, the law may also be used to protect the primary recharge areas of upstate New York.

The State Landfill Law, also sponsored by the Special Legislative Commission on Ground Water Needs for Long Island, is another state-level land use control. The Long Island Land Fill Law, ECL 27.0704, prohibits the siting of new or expansion of existing solid waste municipal landfills within the deep flow recharge areas. It gradually phases out the continued land disposal of municipal solid wastes in existing landfills in the deep flow recharge areas and, eventually, anywhere in Nassau and Suffolk counties.

Connecticut Connecticut's land use control program is closely linked to its ground water classification system. This system, which was implemented in 1980, classified all of the state's land area into one of four ground water classes as described previously in the section on "Classification Systems" earlier in this chapter.

In the Connecticut system, land use activities are controlled by the discharge permit program. For instance, in Classes GAA and GA no industrial waste discharge permits can be issued, which in effect prohibits certain potentially contaminating industries and other land uses. However, not all sources of ground water contamination are addressed through the permit program. For instance, septic tanks and chemical storage tanks are not restricted in any of the land classifications.

The Connecticut system has only been fully operational for about two years and is still undergoing adjustment. It is therefore too early to make a good evaluation of its effectiveness. However, it does appear to have some effective characteristics. From a land use control perspective, it has been used to close some landfills and to exclude siting of other new landfills and discharging industries.

Some of the features of this system that make it effective are as follows:

- Land use controls are used in direct combination with aquifer classification and discharge permitting. It is this combination, not the land use restrictions by themselves, that works well.
- The state had a very strong hydrogeologic data base because of the cooperative program with the USGS, which enabled the classification system to be implemented.
- The state has a strong high-level commitment to ground water protection with capable leadership and supportive resources.
- Connecticut also has an active program to work directly with industry and local communities to understand and use the system properly.

The land use controls of the Connecticut classification system provide a useful tool for protecting ground water, the full implications of which have not yet been fully realized. According to the Department of Environmental Protection (DEP), it has now become the basis for enforcement strategies and priorities. This means that their emphasis will be directed toward phasing out industrial discharges in GA and GB zones.

Another favorable aspect of the Connecticut system is its proven usefulness to industry and local communities in planning efforts. The state DEP now shows local planners how the classification system can be used for land use controls on nonpoint sources through zoning or other controls not available to DEP. Several communities have taken advantage of this assistance.

New Jersey Central Pinelands The Pinelands is a 560-square-mile region (22 percent of New Jersey's land) that is protected by the New Jersey Pinelands Commission as well as the New Jersey Department of Environmental Protection through its Water Quality Standards and Classification System. The critical areas were identified on the basis of ambient water quality and vegetation. Boundaries are drawn on the basis of the location of the drainage areas of the shallow ground-water-fed streams that characterize the area. The discharge limits that are set in this area for nitrate, phosphate, and pH are to protect the environment and are considerably more stringent than public health requirements. Growth is tightly controlled by the commission. No hazardous waste or solid waste treatment, storage, or disposal facilities may be located within the region.

Innovative Approaches Through Property Transfer and Lien Laws

A few states have recently implemented or proposed rather unique and innovative approaches to environmental protection, including protection of ground water, through special laws governing real estate transactions, facility close-downs, and liens on property. Although some aspects of these concepts have been practiced privately for many years on a case-by-case basis, only recently have they been adopted or proposed as general statewide requirements on many types of property sales, transfers, and closures. These approaches are innovative in the sense that they achieve certain environmental protection or enhancement actions through regulations involving property sale and bankruptcy rather than conventional land use restriction, contaminant discharge permitting, and environmental quality standards.

There are two principal approaches included in this category. The first involves requirements to assess, clean up if necessary, and certify that a piece of property is free from major contamination and from threats of future contamination before the property can be sold or closed on. Such certifications have often been required by purchasers of properties as terms and conditions of the sales to relieve the purchaser from assuming potential liabilities associated with the property.

The second approach being implemented or considered by some states involves changing property lien priorities in cases of bankruptcy and related foreclosures, to allow the use of property assets to compensate the state for costs of environmental remedial actions, if needed. Under this special provision, the use of assets can take precedence over other claims such as those of creditors.

Such laws have two major potentially protective effects on ground water. The first is direct benefits realized through required removal or stabilization of contamination sources before the property is transferred. The second potential benefit is the incentives placed on owner/operators, lenders, and investors to minimize future liabilities and losses associated with potential remedial actions, property devaluation, or difficulty in attracting buyers, lenders, investors, or insurers at a later time.

The states of Connecticut and Massachusetts have recently implemented special lien provisions to allow the state to claim assets in bankruptcy cases for remedial action, if needed.

New Jersey's Environmental Cleanup Responsibility Act (ECRA) Explained to the public and industry as being a law analogous to home buyer protection programs, ECRA requires a state declaration that there has been no discharge of hazardous substances or wastes on a property owned by certain industries or that any discharge has or will be cleaned up in accordance with Department of Environmental Protection (DEP) procedures. State certification must come before the sale of industrial establishments or the closure of businesses that involve the use or disposal of toxic or hazardous materials.

The initial step of the ECRA process involves a letter of nonapplicability or an initial notice and Site Evaluations Submission (SES), which must be accomplished no later than 30 days following public release of the decision to close operations or execute an agreement of sale or option to purchase. The SES is essentially an environmental evaluation of the facility. It includes such items as a scaled site map identifying all areas where hazardous substances or wastes are located, a detailed sampling plan, and a decontamination plan if the facility is closing. The sampling plan requires detailed soil, ground water, and surface water monitoring.

The second stage of the ECRA process involves site inspection by agents of the DEP to verify that the information submitted in the SES is complete and correct. If property being reviewed requires cleanup, the DEP reviews, evaluates, and approves a cleanup plan, and a decontamination plan if the facility is closing. Following approval of the company cleanup plan and receipt of the appropriate financial assurances, the DEP authorizes sale, transfer, or closure of the industrial establishment.

The Environmental Cleanup Responsibility Act is a fairly new law, having become effective on December 31, 1983. In its first year and a half, over

1000 transactions were handled. As a result of the ECRA process, 99 cleanups were completed in a 12-month period. While the act is targeted simply at sales and shutdowns, the impact of the law reaches much further and affects owners' land uses. If it is known that ECRA provisions will apply ultimately, certain practices and land uses tend to be avoided. Because new owners and the state can hold previous owners liable if ECRA provisions have not been complied with, contaminated properties cannot simply be dumped by owners using unsound environmental practices. Bankers have become extremely cautious about lending money to borrowers that may be liable under ECRA and have acted to monitor and reinforce ECRA provisions. Failure to follow ECRA prescriptions may involve heavy penalties of as much as $25,000 per day and the potential rescinding of transactions by either the DEP or the purchasee.

Massachusetts "Super Lien" An example of environmentally oriented changes in bankruptcy law is the Massachusetts Priority Lien provision of the Oil and Hazardous Material Cleanup Act, popularly known as "Super Lien." In common bankruptcy law, the order of disposal of assets is state and federal taxes, creditors, and shareholders. The Massachusetts law inserts the state cleanup authority into this order after taxes and before creditors, in order to provide funds for cleanup of contaminated facilities, soils, surface water, and ground water. Although this provides primarily for cleanup, it also provides economic incentive for property investors and creditors (mainly banks and other lending institutions) to assure that those facilities to which the loan or investment is directed take extra care with potentially polluting activities. It is difficult to measure the impact of the law on ground water protection because of the short time in which it has been in place and because of the rather indirect incentives that it provides. It does provide a significant remedial action mechanism that should provide some direct positive contribution to protecting water from degradation.

Conclusions and Recommendations

Land use control is a good complement to source control programs and can significantly increase the level of protection. The land use control programs reviewed were implemented at the county or municipal level (i.e., Massachusetts, New Jersey, New York, Connecticut, and Florida).

If provided with planning and technical support from the state and EPA, many localities can develop effective ground water protection programs employing land use controls. The committee recommends the following:

- **Land use controls should be considered as part of a ground water protection program. Although land use controls are best carried out at the local level,**

state governments can encourage land use controls in combination with other measures to protect ground water.

• The limitation on the effectiveness of land use controls is the degree to which a critical area is developed prior to enactment of the controls. For this reason, land use controls should be implemented at early stages for vulnerable undeveloped areas.

In the areas reviewed by the committee, information collection has been an ongoing aspect of the land use control program. However, the committee believes that it is neither necessary nor possible to have sufficient data to answer all concerns prior to enactment of protective ordinances. These controls can be revised as new data are gathered.

IMPLEMENTATION OF GROUND WATER PROTECTION PROGRAMS

Program Authority, Structure, and Function

Legal Authority and Regulations

Effective state and local ground water protection programs must be based on adequate legal authority and regulations. In the states studied, two distinct bodies of law address ground water resources—state and federal statutory law and state common law. For example, in Connecticut the primary bodies of statutory law on ground water include the state's Clean Water Act, emergency chemical spill statute, hazardous waste statutes, and the Potable Drinking Water Law. In addition, Connecticut's Water Diversion Policy Act regulates withdrawals of ground water through a permit process. Federal statutes that are also applicable to ground water protection include the Resource Conservation and Recovery Act (RCRA), CERCLA ("Superfund") and the Federal Water Pollution Control Act, the Safe Drinking Water Act, the federal Insecticide, Fungicide and Rodenticide Act, and the Surface Mining Act. Common law in most states provides remedies for parties injured by ground water contamination. The primary causes of action under common law include nuisance, negligence, strict liability, and trespass.

Statutory authority should specify the actions that can be taken by state and local programs to protect ground water and to ensure compliance with program requirements. For example, the statutory provisions found in Connecticut laws include the following:

- State and local preconstruction permit requirements.
  - — State regulation of discharges to ground water.

  - State regulation of ground water withdrawals.
  - Local zoning authorities.
- State statutory provisions for abatement of ground water pollution.
  - Scope of resource to be protected and contamination sources covered.
  - Liability for contamination and cleanup.
  - Remedies and penalties for ground water contamination.
- State statutory provisions for supplying potable drinking water.
- State statutory and regulatory provisions concerning nonresidential underground storage of oil and petroleum liquids and chemicals.

The statutory authority provides the basis for development of regulations for program implementation. Each state approaches regulatory development in a manner that is consistent with the enabling legislation, the enforcement philosophy, and the structure of the state agencies involved. Most of the programs examined by the committee had an extensive process for public participation in regulation development. Public participation was seen to be crucial for acceptance of the regulatory approach and the program's implementation. All the programs examined by the committee had specific regulations implementing program requirements. For example, several Cape Cod communities have enacted health regulations governing toxic and hazardous materials management and underground storage of gasoline and other petroleum products. Connecticut, Kansas, New Jersey, and Wisconsin have regulation-based permit programs for discharges to ground and surface water. New Jersey and Wisconsin have water quality standards specified in permitting regulations.

Monitoring and Enforcement

Surveillance and monitoring of the resource to be protected and contamination sources are critical to ensuring compliance with statutory and regulatory goals. Such efforts should seek to achieve the following objectives:

- Assessment of ambient water quality conditions and trends.
- Location and identification of (potential) contamination sources and their impact on water quality.
- Assessment of the impact on water quality and quantity of land use decisions, ground and surface water withdrawals, and other environmental regulatory actions.
- Establishment and modification of water quality standards and permit discharge requirements.
- Assessment of compliance of regulated community with program requirements.
- Evaluation of program effectiveness in meeting goals.

Although it was not possible for the committee to evaluate the surveillance and monitoring efforts of the state and local programs examined, the committee found that strong surveillance and monitoring efforts are critical to ensuring compliance with program requirements. Most of the state and local programs examined have field surveillance and monitoring capability and analytical laboratory support. However, all indicated that such capability falls short of that required to adequately monitor the resource and compliance of the regulated community with program requirements.

Surveillance and monitoring programs must be coupled with strong legal enforcement activities to ensure compliance with program requirements. Such efforts include criminal and civil penalties for violations administered through administrative or judicial procedures, authority to prohibit discharges or construction of polluting activities, and administrative authority to order actions to prevent, cease, or remediate environmental contamination. All the state and local programs examined by the committee have enforcement programs with varying scope. While it was not possible for the committee to evaluate the effectiveness of all state and local enforcement programs examined, several present good examples of the kind of enforcement programs that are needed to protect ground water.

In Kansas, enforcement authority is supplied by the state's health, environmental, and conservation laws. Geohydrologists and engineers of the Kansas Department of Health and Environment (DHE) are stationed in six district offices in addition to the central office staff and are responsible for finding violations and working up cases. Actions are usually brought by the state legal staff of DHE or the Corporation Commission. Maximum daily fines for water contamination per violation are $10,000 to $25,000 per occurrence per day. Both departments can levy fines directly, and appeal to District Court. Violations such as dumping of wastes carry a $5,000 fine plus jail. Fines of $2,500 to $5,000 are automatic on violations such as failure to obtain a permit before starting to drill for a disposal or oil well. The most-used enforcement tool in Kansas is the shutting down of an oil lease or industrial activity. Fines go into a fund that can be used for cleanup. The largest fine for one operator during 1984 was $195,000.

New Jersey and Connecticut, with substantial authority to regulate ground water discharges, have used ground and surface water permit requirements to close down hundreds of municipal solid waste landfills that were contaminating surface and ground water. By the end of 1986, New Jersey expects to have eliminated all landfills owned and operated by or for a single municipality. All solid waste will eventually be managed through regional recycling or resource recovery facilities or disposed of in regional landfills meeting state environmental requirements. These strong enforcement efforts required strong support from the attorney general's office, the state legislature, and the public in both states.

New Jersey's Environmental Cleanup Responsibility Act (ECRA) offers an effective enforcement strategy for ground water protection. ECRA requires certain industrial facility owners to obtain approval from the state Department of Environmental Protection (DEP) before selling property to a new owner. Approval is contingent upon certification that the land is free of hazardous waste contamination or cleanup of any hazardous waste contamination, including contaminated ground water. The DEP and the buyer, through the judicial system, can have the sale voided. Since the program's inception, over 1000 approvals have been sought; 200 sites have required ground water cleanup prior to sale, of which 99 have been completed. This program has resulted in faster and more extensive ground water cleanup than would have occurred through DEP's normal surveillance and enforcement program and has created an incentive for industrial facilities to initiate self-monitoring and cleanup programs to ensure the salability of their property.

Many states have several programs to enforce. These programs may include permits for individual activities or general waste discharge permits covering a variety of activities, or the program may consist of general nondegradation standards. The ability to enforce these programs is heavily dependent on the staff capability of each responsible agency. In New York, pursuant to the New York State Environmental Conservation Law Article 27, and 6 NYCRR 371, the Suffolk County district attorney's office has established an environmental crimes unit. One full-time detective works for this unit. The unit investigates violations of hazardous waste disposal laws, including illegal disposal of hazardous waste. To date, the Suffolk County district attorney's office has obtained six convictions involving illegal disposal of hazardous waste. All are felonies, and at least one jail sentence has been handed out.

Not only are strong enforcement programs important for assuring that program goals are achieved, but enforcement incidents also do much to create favorable public understanding and support. A classic example occurred in 1970, when California officials cited four municipalities and four large industries for long-term pollution violations, all in a one-day hearing. The immediate publicity was helpful, but the long-term support generated as a result has been important to the success of other enforcement actions and has built public confidence in the program. Actions such as this deliver a clear message that violators will be punished. This requires support from the governor's office, the attorney general's office, and the legislature.

Resources and Funding

Properly trained professional staff and supporting equipment are crucial to successful implementation of ground water protection programs. The importance of qualified personnel cannot be overemphasized. They must understand the processes in industries that they are regulating and should follow specific guidelines to correct violations. A review of several court cases shows that relatively few cases are settled in court and that these cases are time-consuming and costly. Staff members need to limit the number of cases going to court and correct violations through on-site visits and the examination of data.

Several of the programs examined by the committee have substantial staffs to carry out their program. For example, in 1984, Kansas spent $10 million in support of ground water permitting and compliance staff support. California staff requirements included 650 positions to handle both surface and ground water quality (48 in the Central Valley region on ground water alone). None of the programs examined, however, had an adequate number of professionally trained staff to carry out the program requirements. State and federal funding for programs falls far short of program needs. Relatively low salaries and civil service requirements often result in high turnover among state and local agency personnel, which hinders long-term effectiveness of the programs.

Conclusions and Recommendations

Successful ground water protection programs require adequate legal authority and substantial funding for planning and design as well as implementation. Other factors affecting the successful implementation of ground water protection programs include the tractability of the problem, the size of the target group whose behavior is to be changed, the extent of behavioral change required, the degree of integration within and among implementing institutions, the amount of media attention directed toward the problem, and the commitment and leadership skills of implementing officials. The committee would have liked to go into further detail concerning the difficulties that may be involved in the eventual implementation of policies designed to protect ground water, but owing to time limitations the members could discuss only a limited number of these issues. Many of the more attractive programs examined, for example, California, Long Island, New York, and Cape Cod, Massachusetts, have benefited from past federal support under Sections 106 and 208 of the Clean Water Act and the Safe Drinking Water Act.

- **The committee recommends that the federal government provide financial support for development and implementation of state- or basin-level programs on the condition that within a specific time period the states are committed to developing self-supporting ground water management programs.**

• Long-term program success requires adequate and continuing funding. This is necessary to maintain a strong regulatory surveillance and enforcement effort with substantial information collection and analytical support. States should consider a variety of funding mechanisms including user and disposal fees as well as general revenues for program support.

• The federal government should also provide technical assistance to state and local governments through research on health and environmental effects of ground water contamination, fate and transport of pollutants, and technologies and strategies for water protection. The federal government should also establish criteria, guidelines, and standards for important ground water contaminants to ensure national consistency and avoid duplication of efforts among states. In addition, the federal government should provide training of state and local officials in ground water management and protection.

Political Mobilization, Support, and Funding

Adoption and implementation of a successful ground water protection program in a state or locality depends largely on a number of intangible factors that can be characterized broadly as a favorable climate for action. An important component of that climate is the existence of a ground water contamination problem. There is enormous variation across the country in the extent to which ground water is actually threatened or contaminated. Some areas, because of geology, are especially vulnerable. Others, because of heavy concentration of industrial activities, are particularly stressed. For instance, in the state of New Jersey a whole series of toxic waste incidents contaminating ground water and the air raised the public awareness, which, in turn, caused the state to develop strong professional environmental protection programs. Public awareness of health problems in this industrial state was high. New Jersey has for many years been well known for having the highest cancer risk in the nation according to National Cancer Institute surveys.

Strong environmental and public health programs often develop in states that have a tradition of concern for human welfare and environmental quality. Still other areas are so dependent on ground water as a source of water supply that possible degradation is especially serious. Objective conditions that suggest that ground water pollution is a serious matter are important in getting such issues on the political agenda but are often by no means sufficient to do so.

Equally or more important than objective conditions are public perceptions, group support, and political leadership. Following is a discussion of the circumstances under which a favorable social and political climate for strong ground water protection may arise.

Discovery of Ground Water Contamination

In the more environmentally aware states, state and local environmental and public health officials have often been the first to discover ground water contamination and recognize its impact on public health. This is frequently the case in states where there has developed environmental health competence particularly with regard to the impact of environmental contamination on human health and a long tradition of protecting drinking water supplies. Noteworthy among these states is California, where this tradition of confronting potential health problems began several decades ago with air pollution episodes and has been reinforced among public health officials attempting to control pesticide poisonings among farm workers.

Individual citizens and citizen groups have, with increasing frequency, become important in exposing potential water contamination and other environmental problems. Although this is particularly the case in states where aggressive environmental or public health agencies have not taken hold, it is true in all states. Citizens are alerted by strange tastes or odors in drinking water or ill health in the community. Concerned citizens have banded together in grass roots organizations that are active in identifying and alleviating sources of pollution and compensating victims. National environmental organizations have encouraged grass roots activities with campaigns to "hunt the dump." It is difficult to assess the true impact of concerned citizens on the identification of potential sources of ground water contamination and the demand for corrective and preventive measures. However, California, Massachusetts, New Jersey, and New York are excellent examples where citizens groups have been particularly active and have played an important role in development of remedial and preventive ground water programs.

The news media have been very important in drawing public attention to pollution events. The treatment that the news media give to cases of ground water pollution appears to vary significantly with the source of discovery. Release of information by governmental officials is often handled differently from problems revealed by pollution victims. For instance, in Tucson, Arizona, the failure of public officials to follow up on potential health effects of trichloroethylene (TCE) discovered in the municipal water supply provided grist for an investigative reporter who responded to citizens' health complaints by doing a house-to-house survey. The media handling of pollution events has an important impact on public confidence in state programs.

On Long Island, the media have been a major factor in the success of the ground water protection program. The program was strongly supported by a major local newspaper. *Newsday* gave in-depth coverage to the 208 plan when it was released and provided excellent coverage on environmental issues. For example, *Newsday*'s revelation that a pharmaceutical company was moving into the deep flow recharge zone led to such public reaction that the developer of the industrial park into which the pharmaceutical company was moving voluntarily placed restrictive covenants on the use of its land to protect ground water. Because of the coverage given to these issues, the developer has found that industries are now attracted to the area because they want to move into no-discharge areas to protect themselves from environmental liability.

Political Culture and Tradition

It is well recognized that different states and regions have different political and social traditions that affect the extent and ways in which government is likely to become involved in regulatory action and the level of government at which action may take place. In New England, for instance, there is a strong tradition of local control with many decisions affecting ground water made through municipalities, many of which are governed by town meetings. Mid-Atlantic states such as New Jersey have a contrasting tradition of powerful state regulatory agencies and commissioners. The Rocky Mountain area in the West has tended to resist governmental regulation of ground water and states and localities have balked at any actions that suggest direct control of land uses that may affect ground water. In some parts of the South, such as in Florida, strong county governments have played an important role in ground water protection. Some states, such as Wisconsin, have long histories of activities in environmental protection. Other states have been considerably less active in governmental regulations aimed at protecting the environment.

Public Awareness

Successful ground water protection programs and their continued support emerge from circumstances where the public is aware of ground water problems and issues. Ground water has been slow in emerging as a major health and environmental issue because (1) the resource is generally invisible to the public; (2) the ways in which ground water is linked to land and other resource uses are complex; and (3) causes and effects of ground water pollution are hard to identify and poorly understood. In order for the ground water issue to take a high-priority position on the public agenda, the issue needs to be perceived as a matter of broad social (rather than narrowly technical) concern.

A history of environmental health programs dating back to the early

1900s provides the basis for many state water protection programs. In Kansas, for example, this historical base for political support was broadened over the years with occasional oil field brine pollution incidents. However, as in several states, it was the major studies under Section 208 of the Clean Water Act in the 1970s that led to the Kansas Ground Water Protection Plan and to revisions of the state's ground water and oil conservation legislation. Legislation designating areas where overpumping threatened the quality of ground water was also passed.

Another state that has a long history of public interest in and awareness of water issues is Wisconsin. Public awareness of potential ground water contamination in Wisconsin seems to arise from the state's long history of concern regarding environmental and ecological issues rather than a series of public health episodes. In Wisconsin, citizens, the media, and many small businesses whose incomes are tied to tourism have long been aware of the impact of water pollution. For instance, the League of Women Voters conducted studies and a series of conferences from 1950 on, concluding that ground and surface water should be regarded as important resources and that protecting the quality of the water was of paramount importance to the future of Wisconsin. Through the involvement and interest of many groups such as this, major water legislation was passed by the mid-1960s that created the Wisconsin Department of Natural Resources (DNR), which has since developed an innovative ground water protection program.

One program that emerged out of the DNR legislation is the Wisconsin Public Intervener Office, which translates understanding and awareness of the water problems into activities that produce compliance with state requirements. It was originally conceived to be a "watchdog" over the new Department of Natural Resources and was established under the auspices of the attorney general. For the past 10 years, the attorney general has appointed two public interveners and supporting clerical staff. In addition, the University of Wisconsin Law School has recently developed a clinical program whereby eight law students work in the public intervener's office for 20 hours a week. The attorney general also established an advisory committee for the office, which now is legally mandated, whose major function is to screen the various requests for action, thereby helping to insulate the public intervener from political pressure. Major new ground water protection legislation was passed in 1984 in which the role of the public intervener was further strengthened by allowing for initiation of lawsuits questioning the constitutionality of laws.

The public intervener handles a variety of subject matters, and private individuals can come to this office for help in dealing with environmental problems. Under the well compensation program, victims can sue the polluter. The intervener reviews and comments on the U.S. Corps of Engineers permit program and participates in hearings on dams and bridges and prepares rules for the Department of Natural Resources or other state agencies. (In Wisconsin, citizens have the constitutional right to petition for rules changes.) Priority areas of concern of the intervener's office in Wisconsin are pesticides, toxic substances, wetland protection, and urban sprawl. Although each year an attempt is made in the legislature to abolish the office, there has always been enough support by those appreciating its effectiveness to retain it. Massachusetts has provisions for citizen suits to mandate governmental action, and the attorney general may intervene on behalf of citizens in any agency regulatory actions.

Recognition by public groups of outstanding achievement by an entity for taking preventive action is another offshoot of public awareness of ground water protection programs. The Governor's Commission on Arizona Environment recently awarded its Certificate of Appreciation to the IBM plant in Tucson, in recognition of its innovative plant design, which prevents any discharges of water or toxic materials.

Education and Technical Assistance

Education can play a key role throughout the ground water program, from its inception to its implementation. One of the impediments to the implementation of effective ground water protection programs is the general shortage on all levels, especially in state governments, of well-trained and experienced hydrogeologists. One function that states can serve in the educational process would be to provide more direct stimulation and support of expanded ground water and related technical academic programs in state-supported universities and colleges. The resultant increase in the numbers and quality of well-trained technical people would have long-term benefits to the general public awareness as well as to program implementation.

In 1982, the Group for the South Fork of eastern Long Island set out to educate both the public and officials in their communities about the impact of population growth and lack of land use planning on water quality and quantity. The group placed ads in local newspapers and published a booklet explaining the technical aspects of the issues and their recommendations for action. Their work resulted in public support in one community for rezoning to larger lot sizes and, in another, the election of a new town council dedicated to protecting the town's environment and drinking water. This has also been the case in other areas on Long Island, where materials for public dissemination ranged from sophisticated technical to broadly educational information on all aspects of water quality and protection. The results have been the development of a strong ground water protection program that is well thought out and implementable.

The need for an ongoing education effort and well-developed avenues for getting information to the public is particularly evident when some new crisis receives attention. Where the public has been kept up to date and involved in decision making for standards and rules of the ground water program, overreaction is often avoided.

Wisconsin is a good example of a state that for many years has expended considerable effort in providing education for its citizens on ground water quality issues. Through the University of Wisconsin's Extension Service, various organizations have kept their members apprised of ground water concerns. For instance, an educational series on ground water management has been the focus of town hall programs conducted on the Education Teleconference Network, which offers listeners an opportunity to ask questions and share ideas.

In recent years, Wisconsin's Department of Natural Resources (DNR) has developed a broad educational program including an annual "Ground Water Report," which is distributed to a large mailing list. It covers ground water issues and related department activities aimed at protecting the quality of ground water resources. The department has published a 32-page magazine supplement on Wisconsin's "Buried Treasure," which has been distributed to over 50,000 people. DNR has developed radio and television public announcements, posters, buttons, and teachers' guides on ground water protection. DNR also has a mobile display that is used at fairs and other public events. In conjunction with the University of Wisconsin, the department also produced a 30-minute film and 15-minute slide show, which have been viewed widely.

The success of Cape Cod communities in developing ground water protection programs has been largely dependent on a strong, ongoing educational and technical assistance effort by the Cape Cod Planning and Economic Development Commission (CCPEDC) and the Association for the Preservation of Cape Cod (APCC). Dating from the inception of the water quality planning process under Section 208 of the federal Clean Water Act in 1975, CCPEDC and APCC have used a variety of educational tools including slide shows, brochures, technical reports, stories in news media, conferences, training programs, and municipal government meetings and hearings to focus public attention on ground water issues. CCPEDC and APCC have brought ground water issues to the attention of all relevant county and municipal bodies such as town meetings, conservation commissions, boards of health, boards of water commissioners, planning boards, and zoning boards of appeal. This effort has been essential to the development and implementation of local by-laws or health regulations dealing with water resource protection districts, underground storage tank regulation, and regulation of storage, use, and disposal of toxic and hazardous materials. This effort has continued to ensure that increased development pressures, changing population dynamics, and turnovers in municipal government personnel do not adversely affect program implementation. On average, CCPDEC personnel attend two to three local meetings per week to educate and/or provide technical assistance to the public or local officials on ground water protection issues.

The land grant college, Kansas State University, and its system of county agricultural agents; the University of Kansas; the state and federal geological surveys; and interim study committees of the Kansas legislature have all contributed to public education and understanding about ground water problems. The water planning studies of the Kansas Department of Health and Environment and those of the Kansas Water Authority, and the public hearings associated with this planning, keep ground water quality issues in the public mind and often provide material for individuals campaigning for election to the state legislature.

Mobilization of Support

Whether strong state or local programs to protect ground water result from the identification of actual or potential pollution problems depends on mobilization of support of experts to provide necessary information, governmental activity to formulate legislation, citizen interest groups that exert pressure, and political leadership. Leadership is a crucial variable in building strong state programs. Successful ground water programs are built upon the foundations of political support.

Citizen interest groups are often crucial in mobilizing support for ground water protection. Local groups have been especially active in promoting a number of ground water protection programs. For example, the Association for the Preservation of Cape Cod worked closely with the Cape Cod Planning and Economic Development Commission in the mid-1970s in support of ground water protection. The two groups used the Section 208 planning monies provided under the Clean Water Act and other government grants to build an information base and a political coalition that pressed for protective action. It is only when water quality and potential damage to ground water resources rank high in importance to the public that sustained attention is likely to occur.

Some citizen groups have formed specifically to deal with ground water contamination or toxic substances problems, while other existing groups have become involved by taking on new missions. Groups that originally form around one issue such as flooding may later become effective supporters of ground water protection. For instance, the Passaic River Basin Committee has been broadly interested in water issues and has added ground water to its list of priorities.

In some cases, however, despite numerous incidents of toxic chemical contamination, the existence of various environmental activists, and a legislative atmosphere receptive to action no effective ground water protection program emerges or is implemented. Often the leadership of a strong personality committed to environmental protection is necessary to catalyze action. This appears to be the case in Wisconsin and New Jersey, where a number of governors have led their states to develop strong environmental protection programs that include innovative ground water protection approaches. In other situations, the leadership of scientists knowledgeable in the area has been important.

Maintaining Momentum

Public and media attention to issues tend to go in cycles; what excites today may bore tomorrow, and what is front page news one year may draw no media attention the next. Because of the cyclical nature of issue attention, it is important for success that ground water protection be institutionalized through the passage of laws and the establishment of organizations to carry out programs. The momentum of programs is mostly likely to be sustained when there is a visible organization, such as the state legislature, that performs an oversight function of the program. Strong bureaucratic skills are necessary to implement programs and to argue successfully for adequate funding and personnel.

At the same time, a strong continuing public education effort is important in carrying out an effective ground water protection program. When there is real awareness of ground water problems and the preventive and remedial program that is or can be carried out to deal with them, one can expect a strong mandate from the public to carry it out. This is translated into political support for strong regulatory and enforcement activities, and for necessary government funding.

Conclusions and Recommendations

Successful ground water protection programs emerge from circumstances where political support is mobilized for the passage and implementation of effective policies. Ground water degradation has been slow in emerging as a major health and environmental issue because the resource is generally invisible to the public, the ways in which it is linked to land and other resource uses are complex, and causes and effects of pollution are hard to identify and poorly understood. In order for the ground water issue to take a high-priority position on the public agenda, the issue needs to be perceived as both a matter of broad social concern and one requiring technical resolution. To facilitate political mobilization, public participation, and support, the committee recommends the following at every governmental level:

- **Decision-making processes concerning ground water should be characterized by openness, should reflect consideration of public attitudes, and should include active participation of public health and environmental interest groups, industry, and the public.**
- **Attention should be directed to the need to attract and develop high-level political leadership to shepherd ground water protection legislation and ensure commitment to continued funding and implementation of ground water programs.**
- **Communication networks must be established and maintained between ground water program managers and the media. Media coverage of ground water issues is more likely to be fair and balanced when managers have established a reputation for openness and accuracy. The scientific community should also share responsibility for assisting in dissemination of clear, accurate, and understandable information by the media.**
- **The sharing and exchange of information regarding ground water protection problems and programs for their resolution should be an ongoing component of every program. This may be achieved through various activities and mechanisms, including regular community meetings, workshops, and symposia that provide full opportunity for discussion, reaction, and recommendation by the interested community concerning the program and issues.**
- **Ongoing educational activities about ground water in the context of environmental protection should be undertaken in the school system at all grade levels.**
- **States should play a key role in expanding the number of well-trained hydrogeologists by providing more support of hydrogeologic programs within state universities and colleges.**
- **A public intervener-type program should be considered when public confidence or interest is not recognized or adequately incorporated in ground water protection programs. A program such as those in Wisconsin and New Jersey can provide the public with an avenue for legal action to address a perceived problem, and at the same time prevent nonmeritorious suits from being filed against an agency. The public intervener should not be subject to political pressure or changes in administrative policy. An independent public advisory group could help to screen the actions to be taken.**

The Role of Economic Analysis

Economic analysis of ground water protection programs is concerned with the identification and measurement of the benefits and costs of protecting the quality of ground water resources. Knowledge of benefits and costs can assist in the formulation of ground water protection programs in at least three ways: (1) it provides a test of the economic feasibility of programs or program elements, (2) it provides criteria for improving the cost-effectiveness of program elements, and (3) it assists in identifying optimal protection programs. Economic analysis has rarely if ever been rigorously used in current efforts to protect ground water. However, such analysis would assure greater benefit from public investments that have been based largely on intuitive reasoning and philosophical commitment.

A ground water resource (such as an aquifer or a portion of an aquifer) is a natural resource with a finite economic value. Its value includes the capitalized net worth of all the services that the resource can be expected to provide over time (use value, such as water for human consumption, water for community uses, water for industrial uses, and dry weather streamflow). The value of each resource service, in turn, is derived from the economic activities that make use of that service (residential uses, commercial activities, industrial production, and recreational uses of streams, for example). In some circumstances, the value of a natural resource may also include intrinsic values, such as option, existence, or bequest values.

Ground water contamination alters the characteristics of the service flows so that they are less valuable now or in the future (e.g., available water is no longer suitable for some uses, such as human consumption, or is suitable only after previously unnecessary treatment). Their capitalized value, the value of the ground water resource, is thereby reduced.

Society may also place a value on the continued existence of uncontaminated ground water resources, regardless of any plans for use of services, during the immediate future (existence value) or extended through future generations (bequest value). Where the potentially contaminated resource is without close substitutes, and the prospective changes are effectively irreversible, even individuals who are not certain to use the resource in the future may attach value to preservation of the opportunity of use (option value). The intended benefits of a ground water protection program are the increased value of the ground water resource, considering both use value and intrinsic value, compared with the total value that would have resulted in the absence of the program. In some cases, benefits may be limited to the cost of restoring a contaminated aquifer in the future (including the value of uses temporarily lost) or by the cost of providing an alternative supply.

Measurement difficulties are frequently encountered in the evaluation of benefits attributable to ground water protection activities. Many current uses are undocumented, complicating the task of estimating future levels with and without the protection program. Even where existing uses are well known, it is often difficult to forecast future uses, as they depend on trends and spatial patterns of economic development in relatively small areas. Where the absence of protection may result in human consumption of contaminated water, only indirect means exist for attributing dollar values to potential health effects. Assigning dollar values to health and ecological impacts or benefits is extremely difficult.

Risk and uncertainty, affecting both analyst and user, are considerations in all aspects of ground water valuation studies. Predictions of future ground water quality, of future ground water uses, and of the value of those uses are characterized by substantial uncertainty. Furthermore, ground water users are uncertain about present and future quality conditions and may actually forego some feasible uses because of perceptions of risk. Thus, uncertainty itself may result in lost benefits, and ground water protection programs that reduce uncertainty may restore those benefits even without changing ground water quality.

Costs of ground water protection programs consist of the value of the resources required to achieve the purposes of the program, including any opportunity costs imposed on those whose use of the resource is precluded or altered in the interest of protection. Unintended external benefits and costs may also result, and they are also a topic of economic analysis, to the extent that they can be anticipated.

Benefits and Costs

Ground water protection programs are presumably undertaken because they are expected to yield benefits at least equal to the costs. If this condition is met, then the total value of all scarce resources, including ground water, is increased as a consequence of the ground water protection program (the program produces positive net benefits). If the positive benefit/cost ratio is not met, then the total value of all resources is decreased by program implementation.

Benefit-cost analysis can also be an issue for individual program elements, as well as for overall protection programs. A specific proposal, such as a plan to regulate and inspect underground gasoline storage tanks, should be capable of providing incremental benefits at least equal to the additional cost of adding that element to an existing program. If this test cannot be met, it is apparent that the overall purpose of the program could be better served in some other way.

When testing program elements, it is important that incremental benefits are not double-counted (the same benefits attributed to more than one program element) and that all incremental costs are considered. Many existing analyses of environmental protection programs confine consideration of costs to outlays by government entities, ignoring cash outlays and opportunity costs borne by resource users. In the case of ground water protection programs, where strategies often include land use restrictions and/or prohibitions on certain activities, private opportunity costs are likely to figure prominently in any comprehensive economic analysis.

Cost-Effectiveness

In addition to having a positive benefit/cost ratio, protection program elements should be designed to accomplish their purposes at the lowest possible social cost. The calculation of cost should consider the full-time stream of future costs, reduced to present value to permit comparison of alternatives. It should also combine costs incurred by public agencies with those borne by resource users and other private parties. Both cash outlays and other types of cost (values of foregone opportunities and reduced asset value, for example) should be reviewed in determining the alternative that places the smallest possible burden on society as a whole.

Optimal Programs

Once program objectives have been stated, economic analysis is helpful in determining which possible program elements are feasible and whether each of these is cost-effective. However, these procedures omit evaluation of the purposes themselves. It is also important to know that the specific goals of the program are the best goals to pursue, that the level of protection sought is the best possible level of protection, and that an appropriate balance has been struck between expenditures for protection and possible later expenditures for cleanup. These considerations are a part of the question of optimality: Is the program proposed the best possible program, all things considered?

Economic analysis does not provide a well-marked pathway to optimality. Rather, the methods of benefit-cost analysis permit the systematic comparison and ranking of alternatives. The result is identification of the best of the alternatives considered, whether that preferred alternative is the best of all possible programs or not. Still, the process of identifying and measuring benefits and costs frequently leads to insights that assist in reformulating programs for improved overall performance.

Comparison of program alternatives should take place after individual elements have been tested for feasibility, all program elements have been designed to be as cost-effective as possible, and efforts have been made to provide the most attractive possible set of alternatives. Comparison of benefits to costs for each of the resulting alternatives leads to a ranking in order of desirability. In most cases, the ranking should be on the basis of the benefit/cost ratio. Where alternatives differ significantly with respect to resulting level of protection, uncertainty associated with benefit and/or cost estimates, fraction of cost borne by private parties, or other factors, it may be desirable to utilize a multiobjective optimizing approach.

Limitations to Analysis

The promotion of ground water protection programs is generally based on the old adage that "an ounce of prevention is worth a pound of cure." However, it is not a simple matter to demonstrate quantitatively the advantages of ground water protection programs. It is true that cleanup efforts for contaminated ground water problems, the development of alternative supplies, and other remedial measures are generally very expensive and are usually only partially effective. However, good prevention programs are also expensive, and for programs now in place effectiveness has not been well demonstrated because of the short period of operation. There are insufficient data at this point to weigh the costs of prevention versus the costs of cleanup in a comprehensive, quantitative manner.

However, remedial action costs run in the billions of dollars as evidenced by Superfund expenditures alone. In addition, major efforts such as state superfund programs and Department of Defense activities for ground water cleanup cost hundreds of millions of dollars per year. On the other hand, the implementation and maintenance of effective ground water protection programs on a statewide basis will probably cost considerably less, on the order of $1 million to $100 million a year depending on the size of the state and the nature of its ground water resource problems. Florida, for instance, currently spends at least $50 million a year implementing its ground water protection program. In addition to statewide programs, there are many costly local programs such as those in southeast Florida and Long Island, New York.

The major difficulty in evaluating the economic value of ground water protection is determining all related costs and benefits. How, when, and to what extent should contaminated ground water be rectified? It may be easy to criticize the pesticide contamination problems such as those in California, Florida, and Long Island, and to suggest that it would have been more economical to have prevented the problem before it occurred; however, at the time of application, it was not widely known that these pesticides would become a problem. Even if the problem had been known, however, there may not have been adequate substitutes available. Therefore, loss of income to the farming industry from banning these pesticides can run into tens of millions of dollars per county.

The conventional wisdom, that virtually any investment in ground water protection is preferable to bearing the cost of later contamination, may generally reflect valid intuition regarding present conditions. However, it does not and cannot, given available methods and data, reflect comprehensive analysis of relative costs and benefits.

Conclusions and Recommendations

Effective ground water protection programs have significant costs associated with them that can, in some cases, exceed the value of the resource or the costs of remedial actions. While analytical techniques are evolving rapidly and data bases are growing, significant application difficulties remain. These difficulties will not be removed until serious attempts are made to perform economic analyses of ground water protection programs and strategies. Meanwhile, social, political, and economic conditions continue to evolve, shifting costs and values so that it is likely to become more and more difficult to strike the right balance between prevention and remedy, or between universal policies and problem-specific measures.

The committee believes that economic analysis is one of the useful ways programs and strategies can be judged. Economic analyses should be conducted of existing and proposed ground water protection measures so that experience can be gained with techniques and data requirements, and decision-makers can become familiar with the results of such analyses. Such analyses have been performed in connection with hazardous waste cleanup activities (assessment of Superfund natural resource damages, for example) and may be useful in evaluating ground water protection programs.

References

Auditor General of California, California State Legislature. 1984. The State Lacks Data Necessary to Determine the Safety of Pesticides. Report P-414. Sacramento.

Carsel, R.F., L.A. Mulkey, M.N. Lorber, and L.B. Baskin. 1985. The pesticide root zone model (PRZM): A procedure for evaluating pesticide leaching threats to groundwater. Ecological Modelling 30:49–69.

Cohen, S. 1985. Office of Pesticide Program, Memo to H. Brass, Office of Drinking Water, U.S. EPA, August 2.

Cohen, D.B., and G.W. Bowes. 1984. Water Quality and Pesticides: A California Risk Assessment Program, Volume 1. State Water Resources Control Board, Sacramento, Calif.

Colorado Department of Health. 1984. Ground-Water Protection in Colorado. Water Quality Control Division, Denver.

Committee on Policy Research Management. Assembly Office of Research. 1985. The Leaching Fields, A Nonpoint Threat to Groundwater. Joint Publications Office, Sacramento, Calif.

Commonwealth of Massachusetts. 1982. Massachusetts Water Supply Protection Atlas. Department of Environmental Quality Engineering, Division of Water Supply, Boston.

Environmental Defense Fund. 1985a. To Burn or Not to Burn: The Economic Advantages of Recycling over Garbage Incineration for New York City. New York.

Environmental Defense Fund. 1985b. No Where To Go: The Universal Failure of Class 1 Hazardous Waste Dump Sites in California. Berkeley, Calif.

Fenneman, N.M. 1946. Physical divisions of the United States: U.S. Geological Survey special map.

Hinkle, M.K. 1985. Conservation vs. Conventional Tillage: Ecological and Environmental Considerations, *in* A System Approach to Conservation Tillage, F.M. D'Itri, ed. Lewis Publishers, Inc.

Holden, P. 1985. Pesticide Contamination of Ground Water in Four States: Issues and Problems. Draft Report to the National Research Council Board on Agriculture, Washington, D.C.

House Committee on Agriculture. 1983. Regulatory Procedures and Public Health Issues in the EPA's Office of Pesticide Programs. Staff report prepared for the Department of Operations, Washington, D.C.

Jury, W.A., W.F. Spencer, and J. J. Farmer. 1983. Behavior assessment model for trace organics in soil, I, Model description. Journal of Environmental Quality 12(4):558–564.

Lacewell, R.D., and S.M. Masud. 1985. Economic and Environmental Implications of IPM, *in* Proceeding of Symposium on Integrated Pest Management. National Research Council. Texas A&M University Press, College Station.

Masud, S.M., and R.D. Lacewell. 1985. Economic Implications of Alternative Cotton IPM Strategies in the United States. Department of Agricultural Economics. Report 85-5. Texas A&M University, College Station.

National Research Council. 1977–1983. Drinking Water and Health, Volumes 1 to 5. National Academy Press, Washington, D.C.

National Research Council. 1985. Reducing Hazardous Waste Generation. National Academy Press, Washington, D.C.

New York State Department of Health. 1981. Report on Ground Water Dependence in New York State. Albany.

New York State Department of Environmental Conservation. 1985. Draft—Upstate New York Groundwater Management Program. Albany.

Pye, V.I., R. Patrick, and J. Quarles. 1983. Groundwater Contamination in the United States. University of Pennsylvania Press, Philadelphia.

Ramlit Associates. 1984. Groundwater Contamination by Pesticides: A California Assessment. Submitted to the State Water Resources Control Board. Berkeley, Calif.

Reynolds, H.T., P.L. Adkisson, and R.F. Smith. 1975. Cotton Insect Pest Management, *in* Introduction to Pest Management, R.L. Metcalf and W.H. Luckman, eds. John Wiley & Sons, New York. 587 pp.

Shoemaker, C.A., and D.W. Onstand. 1983. An optimization analysis of the integration of biological, cultural and chemical control of alfalfa weevil. Environmental Entomology 12:286–295.

State of Connecticut. 1982. A Handbook for Connecticut's Water Quality Standards and Criteria. August 1981. 5th Printing. Department of Environmental Protection, Water Compliance Unit, Hartford.

State of Connecticut. 1985. Map of Water Quality Classifications for the Thames, Southeast Coast, Pawcatuck River Basins. Department of Environmental Protection, Hartford.

U.S. Congress. 1984. Protecting the Nation's Groundwater from Contamination, Volumes I and II. Office of Technology Assessment. U.S. Government Printing Office, Washington, D.C.

U.S. Environmental Protection Agency. 1981. Re-registration Standard Evaluation, p. 8. October 15. Washington, D.C.

U.S. Environmental Protection Agency. 1984. Ground-water Protection Strategy. Office of Ground-water Protection, Washington, D.C.

U.S. Environmental Protection Agency. 1985a. Selected State and Territory Ground-Water Classification Systems. Office of Ground-Water Protection, Washington, D.C.

U.S. Environmental Protection Agency. 1985b. Overview of State Ground-Water Program Summaries. Volume 1. Office of Ground-Water Protection, Washington, D.C.

U.S. Geological Survey. 1982. Water Resources Investigation. Geohydrology of the Meadowbrook Artificial Recharge Site at East Meadow. Report No. 82-4084. Nassau County Department of Public Works, Nassau County, N.Y.

U.S. Geological Survey. 1984. Water Supply Paper 2275. National Water Summary 1984. Hydrologic Event, Selected Water-Quality Trends and Ground-Water Resources. U.S. Government Printing Office, Washington, D.C.

Yoder, D., R. Rodon, R. Usherson, R.C. Johnson, and M.D. Taylor. 1984. Metropolitan Dade County Wellfield Protection Strategies, *in* Proceedings of the National Water Well Association Conference on Ground Water Management. October 29–31, 1984. Orlando, Fla.

Appendixes

APPENDIX A

Bibliography

Arizona Department of Health Services. 1982–1983. Water Quality Assessment. 2005 N. Central Ave., Room 200, Phoenix, Ariz. 85004.

Arizona Department of Health Services, Environmental Health Division. 1983–1986. Groundwater Quality Update. 2005 N. Central Ave., Room 200, Phoenix, Ariz. 85004.

Arizona Department of Health Services. 1984. Arizona State Health Plan (Draft Chapter Environmental Health). 2005 N. Central Ave., Room 200, Phoenix, Ariz. 85004.

Arizona Department of Water Resources. 1985. Management Plan. Phoenix Active Management Area. 99 E. Virginia, Phoenix, Ariz. 85004.

Birch, P.R., and D.V. Jackson. 1979. The industrial potential of materials recovered from municipal solid waste. Conservation and Recycling 3:259.

California Department of Health Services. 1985. Preliminary Report on a Monitoring Program for Organic Chemical Contamination of Large Public Water Systems in California. November. Sacramento.

California Department of Water Resources. 1975. California's ground water. California Department of Water Resources Bulletin 118. 135 pp. Sacramento.

California Department of Water Resources. 1980. Ground water basins in California—A report to the legislature in response to Water Code Section 12924. California Department of Water Resources Bulletin 118-80. 73 pp. Sacramento.

Cape Cod Planning and Economic Development Commission. 1981. Model General Bylaw/Regulation to Control Toxic and Hazardous Materials. Barnstable, Mass.

Cape Cod Planning and Economic Development Commission. 1981. Model Water Resource District Bylaw. Barnstable, Mass.

Cape Cod Planning and Economic Development Commission. 1982. Model Health Regulation to Prevent Leaking of Underground Fuel and Chemical Storage Systems. Barnstable, Mass.

Cape Cod Planning and Economic Development Commission. 1982. Delineating zones of contribution for public supply wells to protect ground water, presented at the National Water Well Association Eastern Regional Conference—Groundwater Management. Oct. 30–Nov. 2, 1983. Orlando, Fla.

Cape Cod Planning and Economic Development Commission. 1982. Water Table Contours and Public Water Supply Well Zone of Contribution. Barnstable, Mass.

Cape Cod Planning and Economic Development Commission. 1984. Results of a regional hazardous waste collection program in Barnstable County, Massachusetts, *in* Hazardous Waste, Volume 1, Number 1, M.A. Liebert, ed. Barnstable, Mass.

Coggins, G., and R.L. Glicksman. 1985. A proposed strategy for prevention of ground water pollution in Kansas. University of Kansas, School of Law, Lawrence.

Cohen, S. 1985. Office of Pesticide Program, Memo to H. Brass, Office of Drinking Water, U.S. EPA, August 2.

Cohen, D.B., and G.W. Bowes. 1984. Water Quality and Pesticides: A California Risk Assessment Program, Volume 1. State Water Resources Control Board, Sacramento.

Cohen, S.Z., S.M. Creeger, R.F. Carsel, and C.G. Enfield. 1984. Potential for pesticide contamination of ground water resulting from agricultural uses, pp. 297–325 *in* Treatment and Disposal of Pesticide Wastes, Krueger and Seiber, eds. American Chemical Society, Washington, D.C.

Commission for Economic Development. 1985. Poisoning Prosperity, The Impacts of Toxics on California's Economy. Office of the Lt. Governor, Sacramento, Calif.

Commonwealth of Massachusetts. 1983. Groundwater Protection Strategy. Department of Environmental Quality Engineering, Division of Water Supply, Boston.

Commonwealth of Massachusetts. 1984. Groundwater Program Summary. Department of Environmental Quality Engineering, Division of Water Supply, Boston.

Commonwealth of Massachusetts. 1984. Listing of Groundwater Protection Controls for Communities Within Massachusetts. Draft Report. Department of Environmental Quality Engineering, Division of Water Supply. Boston.

Erhard, R.F., principal investigator. 1984. Trends in U.S. Groundwater Law, Policy, and Administration. Prepared for Groundwater Task Force of the Edison Electric Institute. Washington, D.C.

Fenneman, N.M. 1946. Physical divisions of the United States: U.S. Geological Survey special map.

Ferris, H. 1984. Probability range in damage predictions as related to sampling decisions. Journal of Nematology 16(3):246–251.

Ferris, H. 1985. Population assessment and management strategies for plant-parasitic nematodes. Ecosystems and Environment 12:285–299.

Florida Department of Environmental Regulation Report. 1982. Florida Ground Water Program Strategy. Prepared by the Ground Water Section Division of Environmental Programs. Tallahassee.

Florida Department of Environmental Regulation Report. 1984. State of Florida Resources Recovery Activity Report. Tallahassee.

Geraghty & Miller, Inc. 1983. Final Report on Technical, Legal and Institutional Assistance. Provided to the Office of Drinking Water, U.S. Environmental Protection Agency. Contract No. 68-01-6425. Syosset, N.Y.

Gibb, J.P., M.J. Barcelona, S.C. Schoock, and M.W. Hampton. 1984. Hazardous Waste in Ogle and Winnebago Counties: Potential Risk via Groundwater Due to Past and Present Activities. Illinois Department of Energy and Natural Resources, Champaign, Ill.

Gordon, W. 1984. A Citizen's Handbook on Groundwater Protection. Natural Resources Defense Council, Inc. New York.

Governor's Commission on Arizona Environment. 1985. Water Quality Fact Sheet. 206 S. 17th Ave., Phoenix, Ariz. 85007.

Healy, H.G. 1981. Estimated pumpage from ground-water sources for public supply and rural domestic use in Florida, 1977. Florida Bureau of Geology Map Series No. 102.

Henderson, T.R., J. Trauberman, and T. Gallagher. 1984. Groundwater Strategies for State Action. Environmental Law Institute, Washington, D.C.

Holden, P. 1985. Pesticide Contamination of Ground Water in Four States: Issues and Problems. Draft Report to the National Research Council Board on Agriculture, Washington, D.C.

Huisingh, D., L. Martin, N. Seldman, and H. Hilger. 1985. Proven Profit from Pollution Prevention. Conference Draft. The Institute for Local Self-Reliance, Washington, D.C.

Illinois Water Plan Task Force. 1984. Strategy for the Protection of Underground Water in Illinois. Special Report No. 8. Illinois Environmental Agency, Springfield.

Kammerer, P.A., Jr. 1981. Ground water-quality atlas of Wisconsin. Wisconsin Geological and Natural History Survey Information Circular No. 39. Madison. 39 pp.

Kansas Department of Health and Environment. 1982. Groundwater Quality Management Plan for the State of Kansas. Bulletin 3-4. Topeka.

Krill, R.M., and W.C. Sonzogni. 1985. Chemical Monitoring of Wisconsin's Groundwater. University of Wisconsin. Madison.

New York State Department of Environmental Conservation. 1983. Executive Summary, Draft—Long Island Groundwater Management Program. Albany.

New York State Department of Environmental Conservation. 1985. Upstate New York Groundwater Management Program Summary. Albany.

New York State Department of Environmental Conservation, Division of Water. 1983. Long Island Groundwater Management Program. Albany.

Powers, J.A., Jr. 1985. An Assessment of U.S. Environmental Protection Agency National Groundwater Strategy on Kansas. Kansas Department of Health and Environment, Topeka.

Radewald, J.D., F. Shibuya, and G.N. McRae. 1985. Biannual multiple drip irrigation applications of chemicals coinciding with nematode population levels leads to a profitable increase in table grape production in the Coachella Valley of Southern California (abstract). Journal of Nematology 17(4):510.

Smith, R.L. 1985. Institutional issues affecting Connecticut's groundwater management program. Paper given at the National Symposium on Institutional Capacity for Groundwater Pollution Control, Denver, Colo., June 20.

State of Connecticut. 1980. Connecticut Water Quality Standards and Criteria. Water Compliance Unit, Department of Environmental Protection, Hartford.

State of Connecticut. 1981. Solid and Hazardous Waste Siting Committee. Connecticut Solid and Hazardous Waste Land Disposal Siting Policy. Hartford.

State of Connecticut. 1982. Water Quality Management Priorities. Water Compliance Unit, Department of Environmental Protection. Hartford.

State of Connecticut. 1984. Protecting Connecticut's Groundwater: A Guide to Groundwater Protection for Local Officials. Department of Environmental Protection, Hartford.

State of Florida. 1984. Groundwater Protection Summary. Department of Environmental Regulation, Tallahassee.

State of Louisiana. 1984. Louisiana Groundwater Program Assessment. Louisiana Department of Environmental Quality, Office of Water Resources, Baton Rouge.

State of New Jersey. 1982. Nonhazardous Waste Regulations, N.J.A.C. 7:26-1 through 6, 14, and 15. New Jersey Department of Environmental Protection, Trenton.

U.S. Congress. 1983. Alternatives for Reducing Hazardous Waste Generation Using End-Product Substitution, Volume 2 (Working Papers), Part D of Technologies and Management Strategies for Hazardous Waste Control. Office of Technology Assessment, Washington, D.C.

University of Arizona, College of Business and Public Administration. 1983. Arizona's Water Resources: Management Options and Obstacles. The Second Annual Policy Forum. Tucson.

Water Currents Newsletter. September 1984. Water Resources Study Commission, Office of Public Works, Department of Transportation and Development. Baton Rouge.

Water, D.D. 1980. Thermal Conversion of Solid Waste and Biomass (An Overview of the Department of Energy Program for the Recovery of Energy and Materials from Urban Solid Waste). Department of Energy ACS Symposium Series 130. Washington, D.C. P. 3.

Wisconsin Legislative Council Staff. 1983. The New Law Relating to Groundwater Management (1983 Wisconsin Act 410). Information Memorandum 84-11. Madison.

APPENDIX **B**

Meeting Participants and Officials Interviewed

STATE AND LOCAL REPRESENTATIVES WHO PARTICIPATED IN COMMITTEE MEETINGS

April 2–3, 1985, Meeting in Washington, D.C.

JOY BARTHOLOMEW, Deputy Secretary, Louisiana Department of Environmental Quality, P.O. Box 44066, Baton Rouge, Louisiana 70804, (504) 342-1265

JOHN GASTON, JR., Division of Water Resources, Department of Environmental Protection, CN-402, Trenton, New Jersey 08625, (609) 984-1787

MARY GEARHART, Colorado Department of Health, 4210 E. 11th Avenue, Denver, Colorado 80220, (303) 320-8333 ext. 3564

JAMES POWER, Bureau of Water Protection, Department of Health and Environment, Building 740, Forbes Field, Topeka, Kansas 66620, (913) 862-9360

HOWARD RHODES, Florida Department of Environmental Regulation, 2600 Blair Stone Road, Tallahassee, Florida 32301, (904) 487-1855

ROBERT SMITH, Water Compliance Unit, Department of Environmental Protection, State of Connecticut, 122 Washington Street, Hartford, Connecticut 06106, (203) 566-2588

May 21–22, 1985, Meeting in Denver, Colorado

JACK B. BALE, Arizona Department of Health Services, Division of Environmental Health Services, 2005 N. Central, Phoenix, Arizona 85004, (602) 257-2338

ROBERT CLARK, Division of Public Water Supplies, Illinois Environmental Protection Agency, 2200 Churchill Road, Springfield, Illinois, (217) 782-9470

BILL DENDY, Bill Dendy & Associates, 429 F Street, Suite 2, Davis, California 95616, (916) 758-3131

DAN HALTON, Bureau of Water Resources, New York State Department of Environmental Conservation, 50 Wolf Road, Albany, New York 12233, (518) 457-3495

KEVIN KESSLER, Wisconsin Department of Natural Resources, Box 7921, Madison, Wisconsin 53707, (608) 267-9350

SARAH MEYLAND, New York State Legislative Commission on Water Resources Needs of Long Island, 43 South Middle Neck Road, Great Neck, New York 10021, (516) 482-7722

SUE NICKERSON, Cape Cod Planning and Economic Development Commission, First District Courthouse, Barnstable, Massachusetts 02630, (617) 362-2511

DARLENE RUIZ, Board Member, State Water Resources Control Board, P.O. Box 100, Sacramento, California 95801, (916) 445-5471

OFFICIALS INTERVIEWED BY COMMITTEE MEMBERS

Arizona

DAVID BARON, Arizona Center for Law in the Public Interest, 32 N. Tucson Boulevard, Tucson, Arizona 85716, (602) 327-9547

LINDY BAUER, Maricopa Association of Governments, 1820 W. Washington, Phoenix, Arizona 85007, (602) 254-6308

PHIL BRIGGS, Arizona Department of Water Resources, 99 E. Virginia, Phoenix, Arizona 85004, (602) 255-1557

WILLIAM CHASE, City of Phoenix, 251 W. Washington, Phoenix, Arizona 85003, (602) 256-3248

VAL DANOS, Arizona Municipal Water Users Association, 1122 E. Washington, Phoenix, Arizona 85034, (602) 256-0999

ROGER FERLAND, Twitt Sievwright & Mills, 2702 N. 3rd Street, Suite 4007, Phoenix, Arizona 85004, (602) 248-9424

SUSAN JO KEITH, Arizona Department of Health Services, 2005 N. Central, Phoenix, Arizona 85004, (602) 255-2350

KATHY KNOX, Arizona State Senate Research, 1700 W. Washington, Phoenix, Arizona 85007, (602) 255-3174

LOIS KULAKOWSKI, League of Women Voters of Arizona, 7541 E. Knollwood Place, Tucson, Arizona 85715, (602) 298-4851

ROBERT MOORE, Agri-Business Council of Arizona, 333 W. Indian School Road, Phoenix, Arizona 85013, (602) 274-3422

PRISCILLA ROBINSON, Southwest Environmental Services, P.O. Box 2231, Tucson, Arizona 85701, (602) 624-2353

KENNETH SCHMIDT, Ground Water Quality Consultant, 4120 N. 20th Street, Suite G, Phoenix, Arizona 85016, (602) 956-8711

ROBERT YOUNGT, Arizona State Land Department, 1624 W. Adams, Phoenix, Arizona 85007, (602) 255-4625

California

DAVID COHEN, Water Resources Control Board, Sacramento, California, (916) 322-8401

HOWARD FERRIS, Professor, Department of Nematology, University of California, Davis, California 92521, (916) 752-6905

ROBERT P. GHIVELLI, Executive Officer, California Regional Water Quality Control Board, 107 South Broadway, Room 4027, Los Angeles, California 90012, (213) 620-4460

LAURIE JOHNSTON, California Department of Food and Agriculture, 1220 N Street, Sacramento, California

MELTON KNIGHT, Chief, Surveillance & Enforcement Branch, Toxic Substances Control Division, California Department of Health Services, 107 South Broadway, Room 7011, Los Angeles, California 90012, (213) 620-2380

REX MAGEE, California Department of Food and Agriculture, 1220 N Street, Sacramento, California, (916) 9280

LAURA MOTT, Natural Resources Defense Council, 25 Keary Street, San Francisco, California 94108, (415) 421-6561

RONALD OSHIMA, California Department of Food and Agriculture, 1220 N Street, Sacramento, California

PETER PRICE, Senior Consultant, Office of Research, Assembly-California Legislature, 1100 J Street, Suite 535, Sacramento, California 95814, (916) 445-1638

DAVID ROE, Senior Attorney, Environmental Defense Fund, 2606 Dwight Way, Berkeley, California 94704, (415) 548-8906

PETER A. ROGERS, Chief, Sanitary Engineering Branch, California Department of Health Services, 714 "P" Street, Room 200, Sacramento, California 95814, (916) 323-6111

DARLENE RUIZ, Board Member, State Water Resources Control Board, P.O. Box 100, Sacramento, California 95801, (916) 445-5471

KEVIN SHAY, Private Consultant, 2609 Capitol Street, Sacramento, California, (916) 441-4075

DAVID SPETH, Water Quality Control Board, Berkeley, California (415) 464-1255

Colorado

RAY CHRISTIANSEN, Colorado Farm Bureau, P.O. Box 5647, Denver, Colorado 80302, (303) 440-4901

SLADE DINGMAN, Peabody Coal Company, 10375 E. Harvard Avenue, Denver, Colorado 80231

CHRISTOPHER F. ERSKINE, Chairman, Colorado Groundwater Protection Strategy Committee, 365 Rangeview Drive, Littleton, Colorado 80120, (303) 798-3272

MARY GEARHART, Colorado Department of Health, 4210 E. 11th Avenue, Denver, Colorado 80220, (303) 320-8333 ext. 3564

ROD KUHARICH, City of Colorado Springs, P.O. Box 1103, Colorado Springs, Colorado 80942

RICHARD PEARL, 120088 W. New Mexico Place, Lakewood, Colorado 80228

BEN SAUNDERS, Colorado High Plains Technical Coordinating Committee, P.O. Box 113, Holyoke, Colorado 80334

Connecticut

TERRY BERTINUSON, (Democrat) State Representative, State Office Building, 165 Capitol Avenue, Hartford, Connecticut 01606

T.J. CASEY, (Republican) State Representative, State Office Building, 165 Capitol Avenue, Hartford, Connecticut 01606

KEN FARONI, Town Planner, Town of Southington

RALPH GOODNO, Executive Director, Housatonic Valley Association, P.O. Box 28, Cornwall Bridge, Connecticut 06754, (203) 672-6670

JEFF HEIDTMAN, Hydrogeologist, Fuss & O'Neill Inc., 210 Main Street, Manchester, Connecticut 06040, (203) 646-2469

JOHN RAABE, Board of Directors, Farmington River Watershed Association, 195 West Main Street, Avon, Connecticut 06001, (203) 678-1241

AUGIE REINHARDT, Plant Engineer, Bostich-Textron, Knollwood Drive, Clinton, Connecticut 06413, (203) 669-8618

WILLIAM RENFRO, Director of Environmental Programs, Northeast Utilities Service Company, 107 Seldon Street, Berlin, Connecticut 06037, (203) 665-5000

Florida

JAMES L. APP, Assistant Dean, Agricultural Program, University of Florida, Gainesville, Florida 32611, (904) 392-1763

JAMES DAVIDSON, Assistant Dean for Research, University of Florida, Gainesville, Florida 32611, (904) 392-1786

RODNEY DEHAN, Florida, Department of Environmental Regulation, 2600 Blairstone Road, Tallahassee, Florida 32301, (904) 487-1855

SAM FLUKER, Professor, Entomology, University of Florida, Gainesville, Florida 32611, (904) 392-4721

GARY KUHL, Executive Director, Southwest Florida Water Management District, 2379 Broad Street, Brooksville, Florida 33512, (904) 796-7211

HOWARD RHODES, Florida Department of Environmental Regulation, 2600 Blair Stone Road, Tallahassee, Florida 32301, (904) 487-1855

DAN SHANKLIN, Agricultural Program, University of Florida, Gainesville, Florida 32611, (904) 392-1901

RICHARD K. SPRENKEL, Associate Professor, Extension IMP, University of Florida, Gainesville, Florida 32611, (904) 627-9236

BOB VAN DEMAN, Director, Solid Waste Management, Pinellas County Resources Recovery Facility, 2800 110th Avenue North, St. Petersburg, Florida 33702, (813) 825-7565

JOE VICK, Assistant to Plant Manager, Government Affairs, Monsanto Fibers and Intermediates Company, P.O. Box 12830, Pensacola, Florida 32575, (904) 968-7109

LEE WORSHAM, Department of Community Affairs Attorney, Legal Section, 2571 Executive Center Circle East, Tallahassee, Florida 32301, (904) 488-0410

DOUG YODER, Assistant Director, Dade County Environmental Resources Management, Room 402, 909 Southeast First Avenue, Miami, Florida 33131

Illinois

C. ROBERT TAYLOR, Department of Agricultural Economics, University of Illinois, Urbana, Illinois 61801, (217) 333-1810

Kansas

JAMES F. AIKEN, Kansas Public Health Association, 2204 West 24th, Wichita, Kansas 67204, (316) 838-1245

RICHARD BLACK, Extension Irrigation Engineer, Kansas State University Cooperative Extension, 231 Seaton Hall, Manhattan, Kansas 66506, (913) 532-6147

STEWART BOONE, Kansas Well Drillers Association, 806 Center Street, P.O. Box 654, Garden City, Kansas 67846, (316) 276-6930

WAYNE BOSSERT, Ground-Water Management District Association, P.O. Box 905, Colby, Kansas 67701, (913) 462-3915

MARY ANN BRADFORD, League of Women Voters, 1809 Webster, Topeka, Kansas 66604, (913) 354-1646

WILLIAM BRYSON, Director, Joint Oil and Gas Regulatory Program, Kansas Department of Health and Environment, Building 740, Forbes Field, Topeka, Kansas 66620, (913) 862-9360

JANICE BUTLER, Kansas Engineering Society, Butler and Associates, P.O. Box 1565, Salina, Kansas 67402, (913) 827-1682

FRED DIEHL, Kansas Municipal Utility League, Box 612, McPherson, Kansas 67460, (316) 241-3314

WILLIAM HAMBELTON, Director, Kansas Geological Survey, 1930 Constant Avenue, Campus West, University of Kansas, Lawrence, Kansas 66045, (913) 864-3965

JOE HARKINS, Director, Kansas Water Office, 109 SW 9th Street, Topeka, Kansas 66612, (913) 296-3185

MARSHA MARSHALL, Kansas Natural Resources Council, 1516 Topeka Avenue, Topeka, Kansas 66612, (913) 233-6707

CHRIS McKINSEY, Counsel, Kansas League of Municipalities, 112 West 7th Street, Topeka, Kansas 66603, (913) 354-9565

DONALD SCHNACKE, Executive Vice President, Kansas Independent Oil and Gas Association, 1400 Merchants National Bank, Topeka, Kansas 66612, (913) 232-7722

SHELBY SMITH, Kansas Chamber of Commerce and Industry, 500 First National Bank, Topeka, Kansas 66603, (913) 357-6321

GERRY STOLTENBERG, Bureau of Water Protection, Kansas Department of Health and Environment, Building 740, Forbes Field, Topeka, Kansas 66620, (913) 862-9360

JANET STUBBS, Home Builders Association, 1217 Merchants National Bank, Topeka, Kansas 66612, (913) 233-9853

Cape Cod, Massachusetts

ARMANDO CARBONELL, Cape Cod Planning and Development Commission, 1st District Court House, Barnstable, Massachusetts 02630, (617) 362-2511

RALPH CIPOLLA, Chairman, Yarmouth Water Quality Committee, 100 Lookout Road, Yarmouth Port, Massachusetts 02675, (617) 362-2578

MICHAEL B. FRIMPTER, Massachusetts Sub-District Chief, U.S. Geological Survey, 150 Causeway Street, Boston, Massachusetts 02115, (617) 223-6816

WILLIAM JOY, Coastal Engineering Company, Inc., Brewster Road, Orleans, Massachusetts 02653, (617) 255-6511

THOMAS MULLEN, Superintendent, Barnstable Fire, Box 540, Barnstable, Massachusetts 02630, (617) 362-6498

STHER SNYDER, Association for the Preservation of Cape Cod, Box 636, Orleans, Massachusetts 02653, (617) 255-4142

WILLIAM TAYLOR, Town Engineer, Rte. 130, Sandwich, Massachusetts 92563, (617) 888-3277

VIRGINIA VELIELA, Chairman, Department of Public Works, Town Hall, Falmouth, Massachusetts 02540, (617) 548-7611

HERB WHITLOCK, Association for the Preservation of Cape Cod, Box 325, Eastham, Massachusetts 92642, (617) 255-0388

New Jersey

ALAN AVERY, Ocean County Planning Board, Court House Annex, Toms River, New Jersey 08753, (201) 929-2054

BART BENNETT, Crummy, DelDeo, Dolan, Griffinger and Vecchione, Gateway 1, Newark, New Jersey 07102, (201) 622-2235

JOE DOUGLASS, Supervisor, Bureau of Industrial Sites Evaluation, 428 E. State Street, Trenton, New Jersey 08625, (609) 633-7141

ELLA FILLAPONE, Passaic River Coalition, 246 Madisonville Road, Basking Ridge, New Jersey 07920, (201) 766-7550

TERRY MOORE, Executive Director, Pinelands Commission, P.O. Box 7, New Lisbon, New Jersey, (609) 984-9344

MAUREEN OGDEN, Assemblywoman, New Jersey State Assembly, 266 Essex Street, Millburn, New Jersey 07041, (201) 467-5153

ROBIN O'MALLEY, Special Assistant to Commissioner Hughey, CN-029, Trenton, New Jersey 08625, (609) 292-2885

ERIC SVENSON, Public Service Electric and Gas, 60 Park Place, Newark, New Jersey 07102, (201) 430-5858

JOHN TRELA, Chief, Bureau of Ground Water Permits, Division of Water Resources, CN 029, Trenton, New Jersey 08618, (609) 292-0424

BOB ZAMPELLA, Pinelands Commission, P.O. Box 7, New Lisbon, New Jersey, (609) 984-9344

New York

ALDO ANDREOLI, Suffolk County Department of Health Services, 225 Raybro Drive, Hauppauge, New York 11787, (516) 348-3782

HAROLD BERGER, Regional Director, New York State Department of Environmental Conservation, Building 40, State University of New York at Stony Brook, Stony Brook, New York 11794, (516) 751-7900

STEVE DRIELACK, Detective Suffolk County District Attorney's Office, 222 Middle Country Road, Suite 222, Smith Town, New York 11787, (516) 360-5300

STEVE ENGELBRIGHT, Museum of Long Island Natural Sciences, Department of Earth and Space Sciences, State University of New York at Stony Brook, Stony Brook, New York 11794, (516) 246-8373

ROBERT FITZPATRICK, Grumman Corporation, Bethpage, New York 11714, (516) 575-3642

D.B. HALTON, New York Department of Environmental Conservation, Bureau of Water Quality Management, 50 Wolf Road, Albany, New York 12233-0011, (518) 457-3656

SARAH MEYLAND, New York Legislative Commission on Water Needs for Long Island, 43 South Middle Neck Road, Great Neck, New York 10021, (516) 482-7722

JIM MULLIGAN, Nassau County Department of Public Works, 1 West Street, Mineola, New York 11501, (516) 535-3905

JOHN OHLMANN, Grumman Corporation, Bethpage, New York 11714, (516) 575-2385

K. ROBERTS, New York Department of Environmental Conservation, Bureau of Water Quality Management, 50 Wolf Road, Albany, New York 12233-0011, (518) 457-3656

EDITH TANNENBAUM, L.I. Regional Planning Board, H. Lee Dennison Building, Veterans Memorial Highway, Hauppauge, New York 11787, (516) 360-5195

MARK WALKER, Center for Environmental Research, Cornell University, Ithaca, New York 14853, (607) 255-7535

Washington, D.C.

MAUREEN HINKEL, National Audubon Society, 645 Pennsylvania Avenue, S.E., Washington, D.C. 20003, (202) 547-9009

MATTHEW LORBER, Office of Pesticides and Toxic Substances, U.S. Environmental Protection Agency, Washington, D.C. 20460

LEE MULKEY, U.S. Environmental Protection Agency, Athens, Georgia, (404) 546-3138

KATHERINE REICHELDERFER, Associate Director, Natural Resource Economics Division, Economic Research Service, U.S. Department of Agriculture, 1301 New York Avenue, N.W., Washington, D.C. 20005-4788, (202) 786-1449

Wisconsin

STEPHEN M. BORN, Professor, Urban and Regional Planning, University of Wisconsin, Madison, Wisconsin 53706, (608) 262-9985 or 262-1004

THOMAS J. DAWSON, Wisconsin Public Intervener, P.O. Box 7857, Madison, Wisconsin 53707-7857, (608) 266-8987

KEVIN KESSLER, Wisconsin Department of Natural Resources, Box 7921, Madison, Wisconsin 53707, (608) 267-9350

JAMES B. MACDONALD, Law School, University of Wisconsin, Madison, Wisconsin 53706, (608) 262-3264

DOUGLAS YANGGEN, Professor of Agricultural Economics, University of Wisconsin, Madison, Wisconsin 53706, (608) 262-3603

APPENDIX **C**

Description of Proposed Regulatory Program for Sources and Activities Potentially Impacting Ground Water Use in Colorado

---

EXPLANATION OF TABLES

TABLE 1—Sources/Activities Subject to Ground Water Quality Regulations Being Developed

This table identifies those sources and activities, or aspects thereof, that are not currently covered with adequate state regulatory programs. Items on this list would be included in a ground water quality control regulation. Two approaches can be considered:

- Immediate coverage utilizing "generic" performance standards, and design criteria.
- Immediate coverage through "performance standards" with design criteria/effluent limitations held in abeyance for rulemaking for specific categories of discharge/activity.*

TABLE 2—Sources/Activities with Variance for Further Study

This table identifies sources and activities that are not currently covered with adequate state regulatory programs, but which do not have an adequate information base on which to base regulatory program coverage. These items would be given a Variance for Further Study and would not be required to meet the ground water quality control program (design criteria/effluent limitations) requirements until such time as they are specifically included by rulemaking. They would be required to meet the "performance standard" element of the regulation. The "further studies" would be conducted on a priority basis as resources allow.

---

*Preferred.

TABLE 3—Sources/Activities Exempted When Equivalent Coverage Is Accomplished by Other State Regulatory Programs

This table identifies sources and activities, or aspects thereof, that would be exempted from ground water quality control regulation requirements due to coverage by some other specifically identified state regulatory program. In order to be included on this exemptions for other coverage list the other regulatory program would need to be found to be "essentially equivalent" with CDH requirements. Such activities/sources would still need to meet the "performance standard" element of the groundwater quality control regulation.

TABLE 4—Sources/Activities Exempted from Coverage

This table identifies those sources and activities which are exempted from coverage under the ground water quality control program (design criteria/effluent limitations) requirements. They would have to meet "performance standards" to the extent that water rights were not materially injured.

TABLE 5—Prohibited Pollutants—To be developed through rulemaking.

This table would list specific pollutants which may not be discharged or placed in subsurface waters, e.g., Dioxins, PCB's, ...

TABLE 6—Restricted Pollutants—To be developed through rulemaking.

This table would list specific pollutants whose presence requires the automatic submittal and approval of a ground water quality protection plan, or its equivalent under other state regulatory programs.

TABLE 1 Sources/Activities Subject to the Ground Water Quality Regulations Being Developed

	Activity/source	Statutory authority	Regulatory program	Approach to protection	Notes	Priority
1.1	Industrial waste-water impoundments	Water Quality Act (25-8-101 et seq.) and/or	Ground Water Quality Control Regulation (to be developed)	Design criteria, Performance standards	Presently no coverage of these sources.	High
		Solid Waste Act (30-20-101 et seq.)	Solid Waste Regulations (6 CCR 1007-2) (being amended)	Design criteria, performance standards	Applies to treatment and disposal, not processing facilities or raw products.	
1.2	Industrial impoundments (processing and raw products)	Water Quality Act	Ground Water Quality Control Regulations	Design criteria, performance standards	Presently no coverage.	High
2.0	Nonhazardous leaks and seepage from storage tanks, processing facilities pipelines, etc.	Water Quality Act (25-8-101 et seq.) and/or	Ground Water Quality Control Regulations (to be developed)	Design criteria, performance standards	State Inspector of Oils, Department of Labor and WQCD have promulgated duplicate regulations for installation and monitoring of underground storage tanks for liquid petroleum products only. However, the regulations do not specifically address the ground water impacts.	High

		Solid Waste Act (30-20-101 et seq.)	Solid Waste Reg's (6 CCR 1007-2) (being amended)	Design criteria, performance standards		
3.0	Land application of secondary domestic wastewater and/or land application for treatment purpose	Water Quality Act (25-1-107 (e))	Ground Water Quality Control Regulations (to be developed)	Design criteria, effluent limitations	CDPS discharge permits being issued on case-by-case basis not specifically pertaining to ground water protection.	High
4.0	Gold and other nonradioactive mining ore processing at custom mills not associated with a mine—i.e., cyanide heap leaches	Water Quality Act (25-8-101 et seq.) and/or	Ground Water Quality Control Regulations (to be developed)	Design criteria, performance standards	Can be applied to processing and disposal aspects.	Medium
		Solid Waste Act (30-20-101 et seq.)	Solid Waste Reg's (6 CCR 1007-2) (being amended)	Design criteria, performance standards	Can be applied disposal aspects.	
5.0	Wastewater sludge disposal	Water Quality Act (25-8-101 et seq.) and/or	Ground Water Quality Control Regulations (to be developed)	Design criteria, performance standards		Medium

(Continued)

TABLE 1—*Continued*

Activity/source	Statutory authority	Regulatory program	Approach to protection	Notes	Priority
	Colorado Health Statutes (25-1-101 et seq.)	Sludge Disposal Reg's (to be developed)	Performance standards	Currently guidelines only. Being proposed for regulation to Colorado Board of Health February, 1984. Application rate calculations to be required.	
6.0 In situ uranium recovery	Water Quality Act (25-8-506)	Ground Water Quality Control Regulations (to be developed)	Design criteria, performance standards	506 regulation needed, since this activity is not under equivalency provision with Radiation Control Act.	Medium
	Radiation Control Act (25-11-101 et seq.)				
7.0 Underground disposal via wells	Water Quality Act (25-8-101 et seq.)	CDPS Regulations for point sources (5 CCR 1002-2)	Effluent limitations, performance standards	Not being implemented at this time though authority exists.	Medium
		or			
		Ground Water Quality Control Regulations (to be developed)	Performance standards, design criteria, and/or effluent limitations	EPA has promulgated proposed regulations for Classes I, III, IV, and V Underground Injection Control, but the final status of the regulations is not known.	

8.0	Domestic/industrial wastewater disposal (surface discharges with subsequent seepage to alluvial aquifers)	Water Quality Act (25-8-101 et seq.)	Ground Water Quality Control Regulations (to be developed); CDPS Regulations for Point Sources (5 CCR 1002-2)	Effluent limitations, performance standards	Current CDPS permit limitations based on receiving stream standards maintenance and do not protect alluvial aquifers from long-term "loading" impacts.	Medium
9.0	Agricultural wastewater impoundments with discharge to ground water	Water Quality Act (25-8-101 et seq.)	Ground Water Quality Control Regulations (to be developed)	Design criteria, performance standards	Primarily feedlots with surface impoundments.	Low
		and/or				
		Solid Waste Act (30-20-101 et seq.)	Solid Waste Reg's (6 CCR 1007-2) (being amended)			
10.0	Domestic sewage lagoons discharging to ground water	Water Quality Act (25-8-101 et seq.)	Ground Water Quality Control Regulation (to be developed)	Design criteria, performance standards	Limited design criteria required for domestic wastewater systems.	Low
		and/or				
		Solid Waste Act (30-20-101 et seq.)	Solid Waste Reg's (6 CCR 1007-2) (being considered)	Design criteria, performance standards	Presently being used on selected cases.	

TABLE 2 Sources/Activities with Variance for Further Study

	Activity/source	Statutory authority	Regulatory program	Remarks	Priority
1.0	Pre-1976 Mined Land Reclamation Act (MLRA) activities with ground water discharge	Water Quality Act (25-8-101 et seq.)	Ground Water Quality Control Regulations (to be developed)	Primarily inactive/abandoned sites for which MLRB authority is lacking. Problem definition and waste categorization analyses under way.	High
		and/or			
		Solid Waste Act (30-20-101 et seq.)	Solid Waste Reg's (6 CCR 1007-2) (being amended)	Certificate of Designation not required but compliance with regulations is necessary. Includes solid as well as liquid wastes.	
2.0	Backflow down wells of pumped irrigation systems of chemicals (pesticides, herbicides insecticides, fungicides, fertilizers, etc.)	Water Quality Act (25-8-101 et seq.)	None at this time	This issue came to light during the June 1983 public meetings as one that is of serious concern in many parts of the state. Poor operational practices or inadequate equipment can allow significant chemical contamination of aquifers. After further problem definition studies by the Department of Agriculture, regulatory options and educational programs should be examined.	High
3.1	Water well construction/abandonment practices	Colorado Water Law (37-90-101 et seq.) and Water Well and Pump Installation Contractors (37-91-101 et seq.)	(2 CCR 402-2)	Interaquifer transmission of low-quality ground water to higher-quality ground water is a potential area of concern due to drilling, construction, and abandonment practices. The legislative declaration is to protect public health through proper practices.	Medium

		Water Quality Act (25-8-101 et seq.)	None at this time		
3.2	Other well-types construction/ abandonment practices	Water Quality Act (25-8-101 et seq.)	None at this time	Further information necessary before regulating drilling practices, but the category includes seismic wells, monitoring wells, etc.	Medium
4.0	Ground Water development/drawdown impacts on quality	Colorado Water Law (37-90-101 et seq.)	(2 CCR 402-2)	Further study needed by DNR, CDH, Dept. of Ag., and others to better describe magnitude and scope of problem. This section is under jurisdiction of the Colorado Ground Water Commission.	Medium
		Water Quality Act (25-8-101 et seq.)	None at this time	CDH authority limited so that material injury to water rights does not occur.	
5.0	Urban runoff—infiltration from detention/retention basins, recharge areas, standpipes, etc.	Water Quality Act (25-8-101 et seq.)	None at this time	Further study needed by CDH and others to identify magnitude and scope of this potential source.	Medium
		Colorado Water Law (37-90-140 et seq.)	(2 CCR 402-2)	Deals with supply, not quality issues.	
6.0	Deicing salts—runoff and piles (highway, runway, etc.)	Water Quality Act (25-8-101 et seq.)	None at this time	Further study needed by CDH and others to better describe magnitude and scope of problem.	Low
7.0	Material stockpile leaching	Water Quality Act (25-8-101 et seq.)	None at this time	Further study needed by CDH and others to better describe magnitude and scope of problem.	Low

(*Continued*)

TABLE 2—*Continued*

	Activity/source	Statutory authority	Regulatory program	Remarks	Priority
8.0	Percolation of atmospheric contaminants—acid rain	Water Quality Act (25-8-101 et seq.)	None at this time	Further study needed by CDH and others to better describe magnitude and scope of problem.	Low
9.0	Construction/excavation impacts to aquifers	Water Quality Act (25-8-101 et seq.)	None at this time	Further study needed by CDH and others to better describe magnitude and scope of problem.	Low
10.0	Geothermal discharges	Water Quality Act (25-8-101 et seq.)	None at this time	Further study needed by CDH and others to better describe magnitude and scope of problem.	Low
11.0	Use or application of pesticides, insecticides, furgicides, etc.	Water Quality Act (25-8-101 et seq.)	None at this time	Further study needed by CDH and others to better describe magnitude and scope of problem.	

TABLE 3 Sources/Activities Exempted When Equivalent Coverage Is Accomplished by Other State Regulatory Programs

	Activity/source	Statutory authority	Regulatory program	Approach to protection	Notes
1.0	Hazardous waste sites	Hazardous Waste Act (25-15-101 et seq.)	Hazardous Waste Reg's (6 CCR 1007-3)*	Design criteria, performance standards	*Siting provisions apply currently, program provisions apply upon EPA delegation.
2.0	Solid waste facilities	Solid Waste Act (30-20-101 et seq.)	Solid Waste Reg's (6 CCR 1007-2)	Design criteria, performance standards	Regulations being updated to reflect HB 1421 requirements for liquids and inactive/closed facilities.
3.0	Open dumps	Solid Waste Act (30-20-101 et seq.)	Solid Waste Reg's (6 CCR 1007-2)	Closure requirements	
4.0	Waste piles	Solid Waste Act	Solid Waste Reg's	Design criteria, performance standards	
5.0	Septic tanks/subsurface percolation systems of domestic wastes	Individual Sewage Disposal Systems Act (25-10-101 et seq.)	ISDS Regulations (5-1003-6)	Design criteria	Regulatory amendments being developed to implement HB 1400 on ISDS systems over 2000 gpd flow.
6.0	Mining facilities—active and inactive (post-1976 MLRA)	Mined Land Reclamation Act (34-32-101 et seq.)	MLRB Regulations (2 CCR 407-2)	Design criteria	Permit and Reclamation Plan Required. Pre-1976 MLRA inactive/abandoned sites not covered.
		Water Quality Act (25-8-101 et seq.)	Colorado Discharge Permit System Reg's (5 CCR 1002-2)	Effluent limitations	Applies to point source discharges.

(*Continued*)

TABLE 3—*Continued*

Activity/source	Statutory authority	Regulatory program	Approach to protection	Notes
	Solid Waste Act (30-20-101 et seq.)	Solid Waste Reg's (6 CCR 1007-2)	Design criteria, performance standards	Certificate of Designation not required. Compliance with reg's required.
7.0 Uranium, vanadium, and other radioactive materials mining and milling facilities				
7.1 Custom mill not associated with a mine	Radiation Control Act (25-11-101 et seq.)	Radiation Control Regulations (6 CCR 1007-1)	Design criteria, effluent limitations, nondegradation of quality	Mill aspects only. Requires equivalency with 25-8-506.
7.2 Mill associated with a mine	Radiation Control Act (25-11-101 et seq.)	Radiation Control Regulations (6 CCR 1007-1)	Design criteria, effluent limitations, nondegradation of quality	Mill aspects only. Requires equivalency under 25-8-506.
	MLR Act (34-32-101 et seq.)	MLRB Reg's (2 CCR 407-1)	Design criteria	For mine aspects of operation.
8.0 Gold and other mining ore processing facilities (nonradioactive)—associated with a mine	MLR Act (34-32-101 et seq.)	MLRB Reg's (2 CCR 407-1)	Design criteria	
9.0 Oil and gas production wastes				

9.1	Injection via Class II wells	Oil and Gas Conservation Act (34-60-106 et seq.)	Oil and Gas Conservation Comm. Reg's (2 CCR 404-1)		Assuming primacy delegation from EPA received.
		Water Quality Act (25-8-101 et seq.)	(5 CCR 1002-2) Reserved section for ground water discharges from point sources	Effluent limitations, performance standards	Not being applied at this time.
9.2	Offsite commercial pits, ponds and lagoons	Solid Waste Act (30-20-101 et seq.)	Solid Waste Reg's (6 CCR 1007-2)	Design criteria, performance standards	Certificate of Designation by county with CDH approval.
		Oil and Gas Conservation Act (34-60-106 et seq.)	Oil and Gas Conservation Commission Reg's (2 CCR 404-1)	Discretionary design criteria	Proposed regulations requiring OGCC permit are out for public review at this time.
10.3	Onsite pits, ponds and lagoons	Oil and Gas Conservation Act (34-60-100 et seq.)	Oil and Gas Conservation Regulations (2 CCR 404-1)	Discretionary design criteria—liners and monitoring	Exemption for these facilities is dependent upon OGCC's adoption of regulations which provide protection equivalent with solid waste regulations and/or ground water quality regulations.
10.4	Land spreading of oil and gas production wastes	Oil and Gas Conservation Act (34-60-100 et seq.)	Oil and Gas Conservation Regulations (2 CCR 404-1)	Prohibited without permit	Exemption for this type of disposal is dependent upon OGCC adopting currently proposed regulations for tracking such wastes from generator to disposal.

(Continued)

TABLE 3—*Continued*

	Activity/source	Statutory authority	Regulatory program	Approach to protection	Notes
10.5	Indiscriminate dumping of oil and gas production wastes	Oil and Gas Conservation Act (34-60-100 et seq.)	Oil and Gas Conservation Regulations (2 CCR 404-1)	Prohibited	Exemption from this type of disposal is dependent upon OGCC adopting currently proposed regulations for tracking such wastes from generator to disposal.
11.0	Spills and indiscriminate dumping of wastes (in sewers, ditches, on land surfaces)	Hazardous Substance Incident Response Act (29-22-101 et seq.)	Not required	Prohibited	
		Radiation Control Act (25-11-101 et seq.)	Radiation Control Regulations (5 CCR 1002-2)		
		Solid Waste Act (30-20-101 et seq.)	Solid Waste Reg's (6 CCR 1007-2)	Design criteria	Discharge of liquid and solid wastes must be in compliance with regulations.
		Water Quality Act (25-8-101 et seq.)	Not required	Prohibited	Discharge to state waters from spills prohibited without a CDPS permit.

		Hazardous Waste Act (25-15-101 et seq.)	Hazardous Waste Reg's (6 CCR 1007-3)	Prohibited	Disposal of hazardous wastes without a permit is prohibited.
		Criminal Code—Hazardous Waste Violation (18-13-112)	Not required	Prohibited	
12.0	Artificial ground water recharge	Colorado Water Law (37-90-101 et seq.) and Water Well and Pump Installation Contractor (37-91-101 et seq.)	(2 CCR 402-2)	Design criteria, contractor licensing	Water quality must not be degraded due to quality of recharge water.

TABLE 4 Sources/Activities Exempted From Coverage

	Activity/source	Authority to regulate	Rationale for exemption
1.0	Surface water storage facilities		Colorado water law prevails so that injury to water rights by administration of Water Quality Act is not allowed. Such facilities are generally a source of good ground water recharge.
2.0	Irrigation return flows	Water Quality Act 25-8-101 et seq.—limited to extent that federal authority requires	Colorado water law prevails so that injury to water rights by administration of Water Quality Act is not allowed.
3.0	Runoff and seepage from agricultural lands	Water Quality Act 25-8-101 et seq.—limited to extent that federal authority requires	Control of this type of nonpoint source in Colorado should be analyzed under the framework of nonpoint source alternatives currently under examination prior to implementation of a potential ground water quality program.

APPENDIX **D**

The New Law Relating to Groundwater Management [1983 Wisconsin Act 410]

This Information Memorandum describes 1983 Wisconsin Act 410, a new law relating to groundwater management. Act 410 became effective on May 11, 1984.

TABLE OF CONTENTS

The Information Memorandum is divided into the following parts:

SOURCE: Reprinted from Information Memorandum 84-11 (dated July 11, 1984) by the Wisconsin Legislative Council Staff, Mark C. Patronsky, Senior Staff Attorney, and Anne Bogar-Rieck, Science Analyst.

I. Funding and Position Authorization
J. Other Provisions

II. ENVIRONMENTAL REPAIR OF SOLID AND HAZARDOUS WASTE DISPOSAL FACILITIES

A. Amendments to the Waste Management Fund
B. Funding and Position Authorization
C. General Purposes of the Environmental Repair Program
D. Environmental Repair Procedures
E. Cost Recovery
F. Other Provisions

III. THE GROUNDWATER FUND

IV. COMPENSATION FOR WELL CONTAMINATION

A. Funding and Position Authorization
B. Submitting Claims
C. Contamination
D. Issuance of Awards; Prorating Awards
E. Limits on Awards
F. Miscellaneous Provisions

V. LABORATORY CERTIFICATION AND REGISTRATION

A. Certification Standards Review Council
B. DHSS and DNR Coordination
C. Laboratory Certification or Registration Required
D. Certification Procedures
E. Registration Procedures
F. Funding

VI. MISCELLANEOUS PROVISIONS

A. Municipalities
B. Department of Transportation
C. Department of Industry, Labor and Human Relations
D. Department of Agriculture, Trade and Consumer Protection
E. Department of Natural Resources
F. Groundwater Coordinating Council
G. Pesticide Review Board
H. University of Wisconsin System
I. Public Intervenor
J. Fox River Management Commission

APPENDIX (1—)HISTORY OF THE SPECIAL COMMITTEE ON GROUNDWATER MANAGEMENT

APPENDIX (2—)LEGISLATIVE ACTION ON 1983 ASSEMBLY BILL 595

ABBREVIATIONS IN THIS MEMORANDUM

Throughout this Memorandum, the following acronyms are used in references to state administrative agencies:

DOA—Department of Administration
DATCP—Department of Agriculture, Trade and Consumer Protection
DHSS—Department of Health and Social Services
DILHR—Department of Industry, Labor and Human Relations
DOJ—Department of Justice
DNR—Department of Natural Resources
DOT—Department of Transportation

REFERENCES IN THIS MEMORANDUM

Each Part of this Memorandum, in addition to the subject matter description, includes references to applicable provisions of Act 410.

Two types of references are included. The first type of reference [e.g., s. 144.44 (4)] refers to statutes which are created or affected by Act 410. The second type of reference [e.g., SECTION 2020] refers to nonstatutory provisions of Act 410.

References to SECTION 2200, terminology changes; SECTION 2201, program responsibility changes; and SECTION 2202, cross-reference changes, are not included in this Memorandum.

Summary of 1983 Wisconsin Act 410

Wisconsin Act 410 contains several major substantive policy initiatives and a variety of appropriations related to groundwater management. The major elements of the Act are:

A. DEVELOPMENT AND IMPLEMENTATION OF GROUNDWATER STANDARDS

1. Department of Natural Resources Sets Standards

a. Act 410 creates a framework for the development and implementation of groundwater protection standards for substances detected in, or with the potential to enter, the groundwater resources of the state.

b. Standards will be set by the DNR for substances of public health concern, based on recommendations of the DHSS, for specific substances which are named in Act 410 and for additional substances identified by state agencies.

c. The DNR will establish, by rule, both an "enforcement standard" and a "preventive action limit" for each substance.

2. Agency Administrative Rules

a. Whenever an enforcement standard and a preventive action limit are set for a substance, each state agency which regulates an activity with the potential to cause groundwater contamination by that substance must review its administrative rules.

b. If existing rules will result in compliance with the groundwater standards, no change in the rules is required.

c. State agencies have considerable latitude in developing administrative rules to comply with the groundwater standards. The rules can be based on performance standards, design specifications or any other means which will achieve compliance.

3. Regulatory Responses When Standards are Exceeded

a. In general, groundwater enforcement standards may not be exceeded, even though the regulated activity complies with all statutes and rules. A regulatory response is always required.

b. The preventive action limit is used as a "warning level" if a substance is detected in groundwater. When a preventive action limit is exceeded, some regulatory response may be necessary.

4. Funding

Act 410 includes an appropriation of $396,300 from the newly-created groundwater fund to enable the DNR, the DATCP and the DHSS to develop and implement groundwater standards.

B. ENVIRONMENTAL REPAIR OF SOLID AND HAZARDOUS WASTE DISPOSAL FACILITIES

1. Purpose of the Environmental Repair Program

a. The purpose of this portion of Act 410 is to create a program for the cleanup of all types of solid and hazardous waste disposal sites, including currently operating facilities, closed facilities and illegal dump sites.

b. This program accomplishes the same purpose as the federal "Superfund," but does so in a significantly different manner.

2. DNR Procedures

a. The DNR is required to compile a list of all places in the state where solid or hazardous waste is currently being disposed or has been disposed in the past.

b. The DNR is then required to create a hazard ranking system and use the hazard ranking system to rank all the sites or facilities on the inventory.

c. Finally, the DNR is required to decide which sites and facilities should be cleaned up, based on the hazard ranking and on several other factors.

3. Funding

a. The costs of site investigation and cleanup are paid by appropriations from the newly-created environmental repair fund.

b. The environmental repair fund receives revenues from fees paid by solid and hazardous waste disposal facilities.

c. To provide initial funding for site investigation and cleanup, Act 410 transfers $554,000 from the newly-created groundwater fund to the environmental repair fund.

4. Cost Recovery

After state funds are expended to clean up a site or facility, costs may be recovered from a person or corporation that meets specified standards of responsibility for the disposal.

C. THE GROUNDWATER FUND

1. Sources of Revenue

Moneys are obtained for the groundwater fund from various fees imposed on activities which have the potential to cause groundwater contamination.

2. Expenditures from the Groundwater Fund

The groundwater fund will be used by several state agencies to develop and implement groundwater standards, to create a groundwater monitoring program and to undertake the cleanup of solid and hazardous waste disposal sites and facilities.

D. COMPENSATION FOR WELL CONTAMINATION

1. When Claims May be Submitted; When Payments Will be Made

a. Claims for compensation may be submitted to the DNR on or after January 1, 1985.

b. The DNR will process claims but withhold payment until after June 30, 1985, when the total amount of claims for the preceding six-month period will be known. If insufficient funds are available to pay all claims in full, the DNR will prorate the awards.

2. Eligibility for Compensation

a. In general, claims may be submitted for contaminated residential wells which serve less than 15 dwelling units and for certain types of livestock wells.

b. Compensation is available only if the level of contamination exceeds health-related standards established either by the federal government or by the state, or if the DNR has issued a written advisory opinion recommending that a specific well not be used.

3. Purposes and Amount of Compensation

a. In general, funds are available for replacement or reconstruction of a contaminated well.

b. Unless claims are prorated, compensation is available for 80% of the cost of repair or replacement of a well, with a limit of $9,600 for an individual grant (representing 80% of $12,000 of eligible costs). If claims are prorated, the percentage will be reduced.

4. Compensation Paid Without Regard to Fault

a. Claims are paid without regard to the cause of the contamination.

b. However, certain acts of the applicant may result in denial of compensation.

5. Funding

Act 410 provides an appropriation of $500,000 from the general fund to pay compensation.

E. LABORATORY CERTIFICATION AND REGISTRATION

1. Purpose of the Certification and Registration Program

Act 410 creates a program in the DNR, with DHSS cooperation, for the certification of commercial testing laboratories and for the registration of laboratories which do not provide commercial testing services but which conduct tests for certain state regulatory programs.

2. Standards for Certification and Registration

To be certified or registered, laboratories must meet standards related to test methodologies, reference to sample test accuracy, quality control and maintenance of records.

F. APPROPRIATIONS AND POSITIONS

1. Total Appropriations and Positions

Act 410 includes a total of $2,436,300 in new appropriations and authorizes 38 new positions for the 1984–85 fiscal year.

2. Summary of Appropriations and Positions

a. Table 1, below, contains a summary of the appropriations and position authorizations in Act 410.

TABLE 1 Summary of Appropriations and Position Authorizations

	1984–85
A. STANDARDS DEVELOPMENT/IMPLEMENTATION	
Natural Resources	
8.0 SEG permanent positions (10/1/84)	$236,700 SEG
Agriculture	
4.0 SEG permanent positions (10/1/84)	101,800 SEG
Health and Social Services	
2.5 SEG permanent positions (10/1/84)	87,800 SEG
B. MONITORING	
Natural Resources	
5.0 SEG permanent positions (10/1/84)	112,600 SEG
5.0 SEG permanent positions (1/1/85)	81,900 SEG
3.0 SEG project positions (10/1/84 to 7/1/87)	57,200 SEG
Monitoring Activities	306,200 SEG
C. LANDFILL CLEANUP PROGRAM	
Natural Resources	
4.0 SEG permanent positions (10/1/84)	104,600 SEG
Site investigations—3 sites	450,000 SEG
D. WELL COMPENSATION PROGRAM	
Natural Resources	
1.5 GPR permanent positions (7/1/84)	47,000 GPR
4.0 GPR project positions (7/1/84)	108,000 GPR
Compensation payments	500,000 GPR
E. UNDERGROUND PETROLEUM STORAGE TANK INVENTORY	
Industry, Labor and Human Relations	
1.0 GPR project position (7/1/84 to 7/1/85)	25,000 GPR
F. SCENIC URBAN WATERWAYS	
Natural Resources	200,000 GPR
G. LABORATORY CERTIFICATION PROGRAM	
Natural Resources	17,500 PR
TOTAL FUNDING:	$1,538,800 SEG
	880,000 GPR
	17,500 PR
POSITION SUMMARY:	31.5 SEG positions
	6.5 GPR positions

SOURCE: Prepared by the Legislative Fiscal Bureau.

b. The appropriations which are described in Table 1 by the term "SEG" are segregated fund revenues, obtained from the newly-created groundwater fund. The purpose of segregated fund revenues in state budgeting is to set aside funds, separate from the state's general fund, for use only for the specific purposes for which the fund is created.

c. The appropriations which are described by the term "GPR" are general purpose revenues. These funds are appropriated by the Legislature from the state's general revenues (largely income and sales tax receipts).

d. The appropriations which are described by the term "PR" are program revenues. These funds are obtained from license fees collected by the state and are retained by the agency to administer the program for which the license is issued.

e. The dates in parentheses are the starting dates and, in some cases, the ending dates of the positions authorized by Act 410.

Part I
The Development and Implementation of Groundwater Standards

This Part of the Memorandum contains a description of the portions of Act 410 relating to the development and implementation of groundwater protection standards. The provisions of Act 410 which relate to the groundwater protection standards procedure include: s. 227.01 (11) (aa), ch. 160 and SECTIONS 2002, 2020, 2025, 2038 (1), (3), (4), (5) and (8) and 2051.

In this Part of the Memorandum, "regulatory agency" means a state agency which regulates activities, facilities or practices that are related to substances which are detected in or have a reasonable probability of entering the groundwater. This term is defined to include the DNR, DATCP and the DOT.

Briefly, Act 410 creates a procedure in new ch. 160, Stats., for the development and implementation of groundwater protection standards for substances detected in, or with the potential to enter, the groundwater resources of the state. Some of these substances are listed in Act 410 and others will be identified by regulatory agencies. Groundwater protection standards are established on a two-tiered basis—both an "enforcement standard" and a "preventive action limit" are determined for each substance. The substances are identified by regulatory agencies which have the authority to regulate activities, practices and facilities which are related to these substances. Once the standards are established, the regulatory agencies are responsible for ensuring compliance with the standards by the activities, practices and facilities which they regulate.

The remainder of this Part of the Memorandum contains a detailed description of the groundwater protection standards procedure in new ch. 160, Stats.

A. ESTABLISHMENT OF ENFORCEMENT STANDARDS

Enforcement standards define when a violation has occurred. When a substance is detected in groundwater in concentrations equal to or greater than its enforcement standard, the activity, practice or facility which is the source of the substance is subject to immediate enforcement action.

The DNR first establishes enforcement standards for substances which are either listed in Act 410 or are submitted to the DNR by a regulatory agency. The DNR determines which of those substances are of public health concern and which are of concern only to the public welfare.

The DHSS is directed to recommend to DNR enforcement standards for substances of public health concern. New ch. 160, Stats., provides a procedure which requires DHSS to recommend certain existing federal standards as enforcement standards, unless a federal standard does not exist for a substance or unless specified conditions are met in determining a standard other than the federal standard. Following the submittal of the DHSS recommendations, the DNR adopts enforcement standards for the substances of public health concern.

For substances which are of public welfare concern only, the DNR alone formulates the enforcement standard. New ch. 160, Stats., creates a procedure which requires DNR to adopt existing federal standards as enforcement standards, unless one does not exist for a substance or unless specified conditions are met in determining a standard other than the federal standard.

Enforcement standards are adopted by the DNR by rule.

B. ESTABLISHMENT OF PREVENTIVE ACTION LIMITS

The preventive action limits for substances function as a "warning level" to determine the need for regulatory responses when a substance is detected in groundwater. Exceeding a preventive action limit creates the possibility that some regulatory response may be necessary. Where a preventive action limit is attained or exceeded, the regulatory agency is required to evaluate the situation and take action necessary to maintain the concentration of the substance at the preventive action limit or at the lowest concentration feasible if the limit has been exceeded. Preventive action limits are intended to provide regulatory agencies with time to take preventive measures to ensure that enforcement standards are not attained or exceeded.

The DNR establishes a preventive action limit for each substance for which an enforcement standard has been established. The preventive action limit is a lesser concentration of the substance, as compared to the enforcement standard. The level of each preventive action limit, in relation to the enforcement standard, is specified by statute based on the health-related characteristics of the particular substance.

Preventive action limits are adopted by the DNR by rule.

C. IDENTIFICATION OF SUBSTANCES

Each regulatory agency is required to submit to the DNR a list of substances (1) which have been detected in, or have a reasonable probability of entering, the groundwater resources of the state and (2) which are related to activities within the agency's authority to regulate. In addition, any person may petition a regulatory agency to include a substance on its list.

The DNR is required to place each substance reported to it into one of three categories to facilitate the prompt establishment of standards for as many substances as possible. The three categories include:

1. Category 1 substances are of highest priority for setting state standards and are those which have been detected in groundwater in concentrations in excess of a federal number.
2. Category 2 substances are those which have been detected in groundwater but not in concentrations in excess of an available federal number.
3. Category 3 substances are of lowest priority and are those which have a "reasonable probability" of being detected in groundwater.

D. REVIEW OF EXISTING REGULATIONS

Upon the promulgation of an enforcement standard for a substance which may affect groundwater quality, each regulatory agency is required to review its rules regarding activities, practices or facilities which are related to that substance. If necessary, regulatory agencies are required to revise their rules so that regulated activities achieve compliance with the requirements of ch. 160, Stats.

Design and management practice rules for facilities, practices and activities must be designed to result in compliance with the preventive action limits, if feasible. Regulatory agencies may not adopt rules defining design and management practice criteria in such a way as to allow an enforcement standard to be attained or exceeded at the "point of standards application." "Point of standards application" is the specific location, depth or distance from an activity at which the concentration of a substance in groundwater is measured for the purpose of determining whether a preventive action limit or an enforcement standard has been attained or exceeded.

Regulatory agencies must review rules for all activities and facilities, except for metallic mining activities which are regulated by the DNR under existing ss. 144.80 to 144.94, Stats. A specific exemption was provided for metallic mining activities because a comprehensive groundwater standards procedure for these activities was recently promulgated [ch. NR 182, Wis. Adm. Code].

A subsequent review of existing rules by a regulatory agency is required whenever a preventive action limit or an enforcement standard is attained or exceeded in the state for a substance related to an activity which that agency regulates. Where a preventive action limit is attained or exceeded, the regulatory agency is directed to revise, if necessary, its rules defining design and management criteria to ensure that compliance with the preventive action limit, where feasible, will be maintained. Where an enforcement standard is attained or exceeded, the regulatory agency is directed to revise its rules, if necessary, to ensure that the enforcement standard is not attained or exceeded in other locations in the future.

E. ADOPTION OF REGULATORY RESPONSES

Each regulatory agency is required to adopt rules setting forth the range of responses which an agency may require of itself or of a person controlling an activity, practice or facility which is the source of a substance, when a preventive action limit or an enforcement standard is attained or exceeded. The intent of this provision is to require only that the regulatory agency identify the type of regulatory responses which will be considered when a preventive action limit or an enforcement standard is attained or exceeded.

F. SITE-SPECIFIC RESPONSES: PREVENTIVE ACTION LIMITS

Where a preventive action limit for a substance is attained or exceeded at some location in the state, those regulatory agencies having jurisdiction over activities which could be the source of the substance are required to assess the cause, evaluate the significance of the concentration and implement appropriate responses. Appropriate responses are those which would achieve the following:

1. Minimize the concentration of the substance at the point of standards application, where technically or economically feasible;
2. Regain and maintain compliance with the preventive action limit or the lowest possible concentration which is technically and economically feasible, if compliance with the preventive action limit is not technically or economically feasible; and
3. Ensure that the enforcement standard is not attained or exceeded at the point of standards application.

The only site-specific response which is subject to statutory guidelines is the prohibition of activities or practices which use or produce the substance. A prohibition on an activity or practice may not be imposed when a preventive action limit is attained or exceeded, unless the regulatory agency:

1. Bases its decision on reliable test data;
2. Determines "to a reasonable certainty, by the greater weight of the credible evidence," that no other remedial response would prevent the "inevitable" violation of the enforcement standard at the point of standards application;
3. Establishes the boundary and duration of the prohibition; and
4. Ensures that any prohibition is related to maintaining compliance with the enforcement standard at the point of standards application.

G. SITE-SPECIFIC RESPONSES: ENFORCEMENT STANDARDS

Regulatory agencies must respond immediately when an enforcement standard is attained or exceeded at any point of standards application. The Act specifies that once compliance with the enforcement standard at a point of standards application is achieved, the regulatory agency continues to be responsible for ensuring compliance with the preventive action limit at the same point of standards application.

If an enforcement standard is attained or exceeded, a regulatory agency must impose a prohibition on activities or practices which use or produce the substance, unless it is demonstrated to the regulatory agency that an alternative response will achieve compliance with the enforcement standard at the point of standards application. A parallel procedure is provided for certain waste disposal facilities [defined as those subject to subch. IV of ch. 144 or ch. 147, Stats.] in which the DNR is directed to require such site-specific remedial actions, including closure, as are necessary to achieve compliance with the enforcement standard at the point of standards application.

H. GROUNDWATER MONITORING

The DNR is required to cooperate with other agencies and the groundwater coordinating council in developing and operating a groundwater monitoring and sampling system. The monitoring system must include, at a minimum, the following five components:

1. Problem assessment monitoring, to detect substances in the groundwater and assess the significance of the concentrations of the detected substances;
2. Regulatory monitoring to determine if preventive action limits or enforcement standards are attained or exceeded;
3. At-risk monitoring, to define and sample at-risk potable wells in areas where substances have been detected in the groundwater;
4. Management practice monitoring, to develop management practices necessary to comply with ch. 160, Stats.; and
5. A plan for managing and coordinating the monitoring program and monitoring information among all regulatory agencies.

The DNR is required to notify a regulatory agency when monitoring data indicate that:

1. A substance has been detected in groundwater;
2. The concentration of a substance in groundwater is changing; or
3. A preventive action limit or enforcement standard has been attained or exceeded at a point of standards application.

The DNR is also required to notify the owner of any potable well and the occupant of any residence served by that well of the results of any monitoring data it obtains from samples of water from that well.

Although the DNR is the lead agency in the monitoring program, other regulatory agencies are not restricted from developing monitoring and sampling programs for groundwater for the purposes of ch. 160, Stats.

I. FUNDING AND POSITION AUTHORIZATION

1. Act 410 appropriates funds and authorizes positions in several agencies for the purposes of developing groundwater standards and implementing the standards. The source of funds for the appropriations is the newly-created groundwater fund. The groundwater fund receives revenues from several different fees, as described in Part III of this Memorandum.

The appropriations are for the 1984–85 fiscal year and the positions are permanent positions. The positions commence on October 1, 1984.

The chart below shows the agencies which receive appropriations for standards development and implementation, the number of positions and the amount of the appropriations.

Agency	Positions	1984–85 Appropriations
DNR	8.0 permanent positions	$236,700
DATCP	4.0 permanent positions	$101,800
DHSS	2.5 permanent positions	$87,800

2. Act 410 appropriates funds and authorizes positions in the DNR for the purposes of the groundwater monitoring program. The source of funds for the appropriations is the newly-created groundwater fund. The groundwater fund receives revenues from several different fees, as described in Part III of this Memorandum.

The appropriations are for the 1984–85 fiscal year. The total appropriation for monitoring is $557,900. Of this amount, $251,700 is for salary, fringe benefits and staff support and $306,200 is for studies and monitoring.

Some of the positions are permanent positions. The remainder of the positions are project positions, which have a limited duration. The chart below shows the DNR positions for the purposes of groundwater monitoring.

Positions	Dates
5.0 permanent positions	Beginning on 10/1/84
5.0 permanent positions	Beginning on 1/1/85
3.0 project positions	10/1/84 to 7/1/87

J. OTHER PROVISIONS

1. The DNR is provided with the authority to petition a regulatory agency for rule-making where:

a. The DNR finds that a preventive action limit or an enforcement standard for a substance is being attained or exceeded at the point of standards application at numerous locations; and

b. The adoption or revision of rules by the regulatory agency is an appropriate response.

Procedures for the petition and the response by the regulatory agency are provided.

2. Regulatory agencies are directed to adopt by rule public participation requirements for the issuance and administrative enforcement of any special order adopted to conform with the requirements of ch. 160, Stats.

Part II

Environmental Repair of Solid and Hazardous Waste Disposal Facilities

This Part of the Memorandum contains a description of those parts of Act 410 relating to environmental repair of solid and hazardous waste disposal facilities.

Provisions in Act 410 relating to this subject are found in ss. 20.370 (2) (cn) and (dr) to (dw), 25.17 (1) (en), 25.46, 70.395 (2) (k), 144.265 (2) (b), 144.435 (1), 144.44 (4) (f) and (g), 144.441, 144.442, 144.76, 144.77 and 227.01 (11) (ac) and SECTIONS 2138 and 2203 (38).

In this Memorandum and in Act 410, the following defined terms are used in relation to the environmental repair of solid and hazardous waste disposal facilities:

"Approved facility" means a solid or hazardous waste disposal facility which has DNR approval of a feasibility report and plan of operation. This term includes an approved mining waste disposal facility.

"Nonapproved facility" means a solid or hazardous waste disposal facility which is licensed by the DNR but does not have feasibility report and plan of operation approval.

"Site or facility" is a broad term which includes approved facilities and non-approved facilities, as well as any other site which is currently used or was used in the past for the purpose of solid or hazardous waste disposal, whether the disposal was legal or illegal.

"Tonnage fee" is the annual fee, calculated in cents per ton of material disposed at the facility, which a solid or hazardous waste disposal facility pays into the waste management fund. The facility only pays the amount by which the tonnage fee exceeds the base fee.

"Base fee" is the annual fee, in a fixed dollar amount, which a solid or hazardous waste disposal facility pays into the waste management fund or the environmental repair fund.

A. AMENDMENTS TO THE WASTE MANAGEMENT FUND

The waste management fund was originally used as a source of revenue for the state's responsibility for long-term care of approved facilities. It also was used to pay for remedial action if the environmental consequences of an occurrence at an approved facility are not anticipated in the plan of operation and pose a substantial hazard to public health or welfare.

The waste management fund is amended by Act 410 so that it is no longer used as a source of revenue to pay for "unanticipated occurrences" at approved facilities. Any environmental repair which is necessary at an approved facility is covered by the newly-created environmental repair fund. As a result of this amendment, the waste management fund will be used as a source of revenue only for the state's responsibility for long-term care of approved facilities after the termination of the owner's 20- or 30-year responsibility for long-term care.

Approved facilities will continue to pay the base fee and the tonnage fee into the waste management fund. The dollar amounts of the tonnage fees established under s. 144.441, Stats., are not changed.

The base fee for nonapproved facilities will be paid into the environmental repair fund instead of the waste management fund. A nonapproved facility will continue to pay tonnage fees into the waste management fund.

B. FUNDING AND POSITION AUTHORIZATION

Act 410 creates a new environmental repair fund. Revenues from several sources are paid into the environmental repair fund and moneys are paid out of the environmental repair fund for remedial action at all types of sites or facilities.

On the effective date of Act 410 (May 11, 1984), the existing funds for hazardous substances spills and abandoned containers are transferred to the environmental repair fund. A separate appropriation is provided within the environmental repair fund for expenditures for each of these purposes.

Under the Act, a total of $554,600 is transferred from the groundwater fund to the environmental repair fund. [A description of the groundwater fund is included in Part III of this Memorandum.] Of this amount, $95,000 is for salaries, fringe benefits and staff support. The Act authorizes four permanent positions for the DNR, beginning on October 1, 1984. The remaining $459,600 is for site investigation and cleanup.

The base fees for nonapproved facilities are paid into the environmental repair fund. The current $100 base fee is retained for nonapproved facilities for which the owner or operator agrees to close the facility on or before July 1, 1999. A new base fee of $1,000 is imposed on a nonapproved facility if the owner or operator does not agree to close the facility on or before July 1, 1999. The owner or operator may qualify for the lower base fee by entering into a closure agreement at any time.

Also, a few nonapproved facilities which accept large amounts of waste will pay a tonnage fee surcharge into the environmental repair fund. The tonnage fee surcharge is calculated as a percentage of the amount of the tonnage fee paid into the waste management fund by a nonapproved facility (i.e., the tonnage fee in excess of the base fee). The surcharge is 25% for any nonapproved facility for which the owner or operator agrees to close the facility on or before July 1, 1999. A 50% surcharge is imposed for any nonapproved facility if the owner or operator does not agree to close the facility on or before July 1, 1999.

Both the new base fee and the surcharges apply to solid or hazardous waste disposed during calendar year 1984. It is expected that a proposal on environmental repair fund fees will be included in the 1985 Budget Bill. It is likely that the proposal will impose a tonnage fee, paid by both approved and nonapproved facilities, to generate revenue for the environmental repair fund.

After environmental repair is undertaken, any funds recovered from responsible persons are returned to the environmental repair fund.

C. GENERAL PURPOSES OF THE ENVIRONMENTAL REPAIR PROGRAM

The overall purpose of the environmental repair program is to establish a mechanism for inventory, investigation, hazard ranking and environmental repair of any type of site or facility, whether or not the site or facility is licensed and whether or not the site or facility is in current operation. Environmental repair can be undertaken for any occurrence at a site or facility which presents a substantial danger to public health or welfare or the environment.

The environmental repair fund may be used as a source of the state's cost share for federal superfund projects. Also, the environmental repair fund is a source of funds for the hazardous substances spills program and the abandoned container program.

D. ENVIRONMENTAL REPAIR PROCEDURES

The DNR is given authority to undertake investigation and analysis at sites or facilities or to enter into contracts for this purpose. The investigation and analysis are for the purpose of determining the existence and extent of environmental pollution and determining the degree of danger to public health or welfare or the environment. There are three steps in the environmental repair procedure:

1. Step one in the environmental repair procedure is the preparation by the DNR of an inventory of sites or facilities which cause or threaten to cause environmental pollution. The inventory will be comprehensive and will not be based on the potential degree of harm associated with the site or facility. The DNR is required to publish the inventory and any amendments to it in a major newspaper with statewide circulation. The inventory is not promulgated as a rule. The DNR is not required to conduct a public hearing when the inventory is published and the inventory is not subject to judicial review.
2. Step two in the environmental repair procedure is the preparation by the DNR of a hazard ranking list. The DNR is required to promulgate hazard ranking criteria by rule. These criteria will include a variety of factors regarding the degree of risk or potential for harm associated with the facility and the urgency of taking remedial action. The hazard ranking list will rank sites or facilities, with the order determined by the hazard ranking criteria. The DNR is required to publish the hazard ranking list and any amendments to it in a major newspaper with statewide circulation. The hazard ranking list is not promulgated as a rule. Upon request, the DNR is required to hold a public informational hearing on the hazard ranking list, but the DNR is not required to conduct a hearing as a contested case. The hazard ranking is not subject to judicial review.
3. Step three in the environmental repair procedure is the environmental repair work. The DNR has the authority to undertake environmental repair on its own, to enter into contracts for environmental repair or to enter into agreements whereby parties responsible for the disposal site will undertake environmental repair.

In determining the order in which remedial action will be undertaken at sites or facilities, the DNR is authorized to consider a variety of factors, including: (1) the hazard ranking of the site or facility; (2) the amount of funds available for environmental repair; (3) the information available about a site or facility; (4) the willingness and ability of a responsible person to undertake or assist in remedial action; and (5) the availability of federal funds. The DNR will not consider the amount of money paid into the environmental repair fund by a facility in determining whether to undertake remedial action.

When the DNR makes a decision to undertake remedial action at a site or facility, any person may request a public hearing on the subject of whether the proposed expenditure is within the purposes of the environmental repair program and is reasonable in relation to the cost of obtaining similar materials and services. The DNR is not required to treat the hearing as a contested case. The decision of the DNR to undertake remedial action is subject to judicial review.

E. COST RECOVERY

Act 410 provides an opportunity to recover costs from persons responsible for the disposal after moneys have been expended from the environmental repair fund. The cost recovery provisions apply whether a site or facility is currently in operation or is no longer in use. Cost recovery is obtained in a lawsuit filed by the Attorney General.

An owner or operator of a site or facility is "responsible," and cost recovery is available, if the person knew or should have known at the time the disposal occurred that the disposal would be likely to result in the release of a substance so as to cause a substantial danger to public health or welfare or the environment. Any person, including an owner or operator, is "responsible" if:

1. The person violated a statute, rule, plan approval or order in effect at the time the disposal occurred and the violation caused or contributed to the condition at the site or facility; or
2. The person's action related to the disposal caused or contributed to the condition of the site or facility and would result in liability under common law at the time the disposal occurred, based on existing knowledge and standards of conduct for that person at the time the disposal occurred.

The DNR is given authority to accept in-kind payments of environmental repair work from responsible persons. In connection with the in-kind payment, the DNR is also authorized to enter into agreements to limit cost recovery or to waive part or all of the liability of the responsible person.

Two exceptions are provided for cost recovery: (1) no cost recovery is available for environmental repair work connected with releases of substances in compliance with a permit; and (2) no cost recovery is available from purchasers of land who had no actual knowledge and no reason to know of the existence of a site or facility when the land was purchased.

F. OTHER PROVISIONS

1. The DNR is authorized to undertake emergency action under the environmental repair program. This allows the DNR to bypass the considerations for determining the order of remedial action. The DNR is not required to hold a hearing on a decision to undertake emergency action but the DNR's decision is subject to judicial review.
2. The DNR is authorized to enter property at reasonable times and upon notice to the owner or occupant to take action under the environmental repair program. Notice is not required if delay may result in an imminent risk to public health or safety or the environment.

3. The mining investment and local impact fund, which consists of moneys obtained from the net proceeds tax on metallic mineral mining, can be used as a source of funds for environmental repair at approved mining waste facilities.

4. The DNR is required to promulgate rules establishing criteria for decisions on the amount of state funds to be expended for the required cost share at federal superfund sites.

5. The DNR is authorized to require monitoring at nonapproved facilities as a condition of licensing. A nonapproved facility is entitled to a credit against the $1,000 base fee equal to the cost of monitoring, with a maximum credit of $900.

6. The DNR is authorized to issue special orders requiring a person, who was the owner or operator of a facility which is no longer in operation, to conduct monitoring at the facility. If a municipality is responsible for monitoring at a closed facility, the municipality is only required to pay for monitoring costs up to $3 per resident. The environmental repair fund will pay the additional monitoring costs at municipal facilities.

7. New statutory language provides authorization for an abandoned container program. Although the DNR previously had an appropriation for this purpose, there was no statutory language relating to abandoned containers.

Part III
The Groundwater Fund

This Part of the Memorandum contains a summary, in Table 2, of the "groundwater fees" imposed annually under Act 410. All of the groundwater fees shown in Table 2 are paid into the groundwater fund, which is created by Act 410. The dollar amounts of the groundwater fees are specified in Act 410. The amount of the groundwater fee collected from each activity is an estimate, based on recent information about each activity. Table 2 also lists the "basic fee," if any, for the same activities. The basic fee is the existing fee for each license, permit, inspection or other activity. All groundwater fees imposed by Act 410 are in addition to the basic fees.

The groundwater fees are deposited in the groundwater fund. The groundwater fund is a segregated fund, which means that the fund is set apart in an account which is separate from the state's general fund. The groundwater fund is available only to pay for setting groundwater standards, establishing a groundwater monitoring program and undertaking inventory and site investigation of abandoned landfills.

The specific appropriations from the groundwater fund are described in Part I, Section I, and Part II, Section B [ss. 25.17 (1) (gm) and 25.48].

TABLE 2 Summary of Fees Imposed Annually Under Act 410

A. DEPARTMENT OF AGRICULTURE, TRADE AND CONSUMER PROTECTION

1. Pesticides [s. 94.681 (2)]

Basic fee: $100 license fee
Groundwater fee: $2,000 for each manufacturer of the active ingredients of a pesticide [40 manufacturers]
$300 for distributors [660 distributors]
$100 for distributors of only one pesticide [100 distributors]
Total groundwater fee collected: $ 288,000
[Fee collected before the end of the year for an annual license issued for the following calendar year. Fee first collected in late 1984 for the 1985 licenses.]

2. Fertilizer [s. 94.64 (4) (an)]

Basic fee: $0.10 per ton for inspections; $0.10 per ton for research
Groundwater fee: $0.10 per ton; 1,300,000 tons
Total groundwater fee collected: $ 130,000
[Fee collected after each six-month reporting period—January to June and July to December. Fee first collected for the July to December 1984 reporting period.]

B. DEPARTMENT OF NATURAL RESOURCES

1. Septage Haulers [s. 146.20 (4s) (d)]

Basic fee: $25 license fee for residents; $50 for nonresidents
Groundwater fee: $50 per license; 705 licenses
Total groundwater fee collected: $ 35,300
[Fee collected in June for the following fiscal year. Fee first collected in June 1984.]

2. Wastewater and Sludge Land Disposal [s. 147.033]

Basic fee: None
Groundwater fee: $100 per permittee; 1,010 permittees
Total groundwater fee collected: $ 101,000
[Fee proposed by DNR to be collected before the end of the year for the subsequent calendar year. Fee proposed to be first collected for calendar year 1985.]

3. Solid and Hazardous Waste Disposal [s. 144.441 (7) and SEC. 2203 (38)]

Basic fee: Generally, $0.035 per ton for solid waste; $0.35 per ton for hazardous waste; $0.001 for mine waste rock
Within this range, other fees apply to different wastes
Groundwater fee: $0.10 per ton; 6,300,000 tons
$0.01 per ton for prospecting or mining waste; no current disposal
Total groundwater fee collected: $ 630,000
[Fee collected in April for the total amount of waste disposed during the previous calendar year. Fee first collected for waste disposed on and after January 1, 1984.]

C. DEPARTMENT OF INDUSTRY, LABOR AND HUMAN RELATIONS

1. Private Sewage Systems [s. 145.19 (6) and SEC. 2203 (25)]

Basic fee: Basic sanitary permit fee is $41; counties may charge more
Groundwater fee: $25 per permit; 14,000 permits
Total groundwater fee collected: $ 350,000
[Fee collected whenever a county issues a sanitary permit. Fee first collected for permits issued on July 1, 1984.]

2. Petroleum Product Storage Tanks [s. 101.14 (5)]

Basic fee: $32 or $43 for plan examination, depending on size of tank; $43 for site inspection
Groundwater fee: $100 per inspection; 425 inspections
Total groundwater fee collected: $ 42,500
[Fee collected when an application for plan approval is submitted to the DILHR. Fee first collected for plan approvals submitted on May 11, 1984.]

TOTAL AMOUNT COLLECTED ANNUALLY
FOR THE GROUNDWATER FUND: $1,576,800

SOURCE: Prepared by Legislative Council Staff.

Part IV
Compensation for Well Contamination

This Part of the Memorandum contains a description of the compensation program for well contamination. The program pays compensation on a no-fault basis for residential or livestock wells which are contaminated by substances in excess of the health-related groundwater standards for those substances.

Provisions in Act 410 related to the compensation program are found in ss. 20.370 (2) (eb) and (ec), 71.03 (2) (g) and 144.027 and SECTIONS 2033 and 2203 (45).

A. FUNDING AND POSITION AUTHORIZATION

Act 410 provides an appropriation of $155,000 to fund 1.5 permanent positions and 4.0 project positions which begin on May 11, 1984, for the purpose of administering the program. The Act does not specify an ending date for the project positions. Under s. 230.27, a project position may not be continued for more than four years.

Act 410 provides an appropriation of $500,000 in general purpose revenues (GPR) to pay compensation for claims submitted from January 1 to June 30, 1985.

B. SUBMITTING CLAIMS

The compensation program is administered by the DNR. The DNR will promulgate rules to implement the program, receive applications for claims and decide whether to pay compensation.

The compensation program has a delayed effective date. Claims for compensation may not be submitted until January 1, 1985. The delayed effective date is intended to permit the DNR to develop rules to implement the compensation program.

A claim may be submitted to the DNR for a private water supply which is contaminated. "Private water supply" includes any of the following types of wells:

1. A residential well which is used as a source of potable water for humans or humans and livestock and is connected to 14 or less dwelling units.
2. A livestock well which complies with Grade A milk production standards.
3. A livestock well which does not comply with Grade A standards if the well is constructed by boring or drilling.

An individual, partnership, corporation or association which is the landowner or lessee of the property where a contaminated well is located may submit a claim. A claim may not be submitted by the state or any of its political subdivisions, a state agency, the federal government or an interstate agency.

The DNR determines if the claim is complete. Submission of a claim constitutes consent to allow the DNR to inspect the property at reasonable times and consent by the claimant to cooperate in state enforcement actions against any person or activity alleged to have caused the contamination.

C. CONTAMINATION

A private water supply is contaminated and eligible for compensation if it produces water containing substances of public health concern (1) in excess of a federal primary maximum contaminant level or (2) in excess of one of the new state enforcement standards developed by rule under ch. 160, Stats. The claimant must obtain and pay for two tests necessary to prove that the well is contaminated.

In addition, a person may submit a claim if the DNR issues a written advisory opinion recommending that the well not be used because of potential human health risks.

Compensation is available even if the well is contaminated before the effective date of Act 410 (May 11, 1984), provided that the contamination continues when the claim is submitted.

D. ISSUANCE OF AWARDS; PRORATING AWARDS

If the DNR determines that the well is contaminated and the claimant has met all requirements of the compensation program, the DNR is required to issue an award for the following purposes:

1. Testing the well.
2. Obtaining an alternate water supply (i.e., bottled water) for up to a one-year period.
3. If the contamination is expected to continue for more than one year:

   a. Reconstructing the well or constructing a new well, or connecting the premises to a private or public water supply.
   b. Purchasing a new pump, if a larger pump is necessary due to the greater depth of a new or reconstructed well.
   c. Abandoning the contaminated well.
   d. Purchasing water treatment equipment as a last resort, if reconstruction or replacement of the well will not remedy the contamination, and connection to another water supply is not feasible.

Awards are issued on a no-fault basis. Contributory negligence of the claimant is not a bar to recovery.

The DNR is required to aggregate claims made between January 1, 1985 and June 30, 1985 and issue all awards after June 30, 1985. The DNR will prorate payments based on the amount of funds available for compensation, with a maximum state share of 80% of the eligible costs. The DNR will adjust the percentage of the state share in order to prorate the claims. A ceiling of $12,000 is placed on the amount of eligible costs per claim.

The claimant is required to make a copayment equal to the remainder of the eligible costs which are not paid by the DNR, with a minimum copayment of $250.

E. LIMITS ON AWARDS

The following limits apply to awards under the compensation program:

1. All awards are based on the "usual and customary costs" of the remedial work, as determined by rule by the DNR.
2. If a livestock well is contaminated only by nitrates, an award is available only if the well produces water containing nitrates in excess of 40 parts per million expressed as nitrate-nitrogen.
3. If an award is made for reconstruction or replacement of a well, the well must be constructed by a well driller licensed under ch. 162, Stats.
4. If the Secretary of the DNR determines that a remedial response to groundwater contamination by a regulatory agency can be expected to remedy the contamination in a well in less than two years, the Secretary may delay awards for water treatment, well reconstruction, well replacement or connec-

tion to another water supply for up to a two-year period. The DNR must issue an award for bottled water in the interim.

5. If the contaminated well is a sand point well, the DNR may issue an award for replacement of the sand point well with a drilled well only if: (a) replacement by another sand point is not feasible; (b) the DNR determines that the claimant had no reason to believe that the sand point well would become contaminated when it was constructed; and (c) the well will serve a principal residence.

F. MISCELLANEOUS PROVISIONS

The well compensation program includes several miscellaneous provisions:

1. Amounts received as compensation are exempt from state income taxation.

2. After receiving a payment under the compensation program, a person may not receive another payment for the same parcel during the subsequent 10-year period. However, if the claimant receives an award for a new or reconstructed well or for connection to another private water supply and if the contamination is not remedied by the work done, the claimant is eligible to receive one additional award in the subsequent 10-year period.

3. The claimant may use the claims process in lieu of, or in addition to, any civil remedies which may be available. The claimant is not required to submit a claim before pursuing a civil remedy. The statute of limitations for a civil action is suspended while the claim is being processed, to ensure that the claimant's civil remedies are not lost. The findings and conclusions of the DNR in connection with issuing an award are not admissible in a civil action.

4. The state is subrogated to the rights of the claimant, so that: (a) a claimant who receives compensation for the same purpose from another source must repay the groundwater contamination fund; and (b) the state can recover money paid out of the groundwater contamination fund by suing the alleged polluter.

5. The DNR is required to deny a claim if the contaminant was introduced into the well through the plumbing connected to the well, the contamination was caused intentionally by the claimant or the claimant submitted a fraudulent claim.

6. A person who, in order to obtain an award, intentionally contaminates a well or submits a fraudulent application is subject to a forfeiture of $100 to $1,000 and is required to repay the award.

7. An individual who can obtain compensation under one of the two mining compensation statutes must do so, rather than use the new compensation program.

8. Claims may not be submitted under the compensation program after January 1, 1987. The Legislative Council is requested to study the provision of private water supplies which are contaminated and to report to the Legislature by July 1, 1986.

Part V
Laboratory Certification and Registration

This Part of the Memorandum contains a description of the portions of Act 410 relating to laboratory certification.

The provisions of Act 410 which relate to the laboratory certification and registration procedures are as follows: ss. 15.107 (11), 20.370 (3) (dj), 143.15 (5), (8) and (9) and 144.95 and SECTION 2001.

Briefly, Act 410 provides guidelines on laboratory certification and registration procedures to be established by rule by the DNR. Act 410 also creates a certification standards review council, provides for cooperation by DNR with the DHSS's laboratory certification program and states when laboratory certification or registration may be required.

The remainder of this Part of the Memorandum contains a description of the laboratory certification and registration procedures.

A. CERTIFICATION STANDARDS REVIEW COUNCIL

A certification standards review council is created in the DOA. The council consists of eight members appointed by the Secretary of the DOA, representing municipalities, industrial laboratories, commercial laboratories, public water utilities, solid and hazardous waste disposal facilities and agricultural interests. A ninth member is appointed by the Chancellor of the University of Wisconsin-Madison to represent the State Laboratory of Hygiene.

The council is directed to review the DNR laboratory certification and registration program and to make recommendations to the DNR regarding test categories, reference sample testing, standards for certification and registration and other aspects of the program.

B. DHSS AND DNR COORDINATION

Act 410 contains three provisions to facilitate cooperation between the DNR and DHSS regarding their respective laboratory certification and registration programs as follows:

1. Rules emanating from either Department relating to laboratory certification must be approved by both Departments;
2. The Departments must enter into a memorandum of understanding regarding the responsibilities of each Department in administering its laboratory certification programs;
3. Each Department must accept the certification or registration of a laboratory by the other Department; and
4. Both the DNR and the DHSS must submit rules developed by those agencies to the State Laboratory of Hygiene for review and comment.

C. LABORATORY CERTIFICATION OR REGISTRATION REQUIRED

The DNR may impose certification and registration requirements if the results of tests are required to be submitted to the DNR in a "covered program." A "covered program" is one of the following elements of specific regulatory programs:

1. A feasibility report, plan of operation or condition of any license issued for a solid or hazardous waste facility;
2. An application for a mining permit;
3. Monitoring required under a ch. 147, Stats., permit;
4. The replacement of a well or provision of alternative water supplies;
5. Groundwater monitoring;
6. The management or enforcement of the safe drinking water program;
7. The terms of DNR contracts;
8. An investigation of a hazardous substance discharge; and
9. A regulatory program specified by the DNR by rule.

The DNR may require by rule that tests to be submitted to the DNR under a covered program must be conducted by a laboratory which is certified or registered to conduct tests in a specific test category. The DNR may also require that tests must be conducted by a certified laboratory in the event that requirements for registration do not meet the requirements of an applicable federal law.

The DNR may recognize the certification, registration, licensure or approval of a laboratory by a private organization, another state or a federal agency, if the standards for certification, registration, licensure or approval comply with DNR laboratory certification and registration standards.

D. CERTIFICATION PROCEDURES

Laboratory certification procedures are provided for laboratories performing tests for hire in connection with a covered program. The DNR must establish by rule uniform minimum criteria to be used to evaluate laboratories for certification. Specifically, the DNR establishes criteria for the following:

1. An accepted methodology for conducting tests in each test category;
2. Reference sample test accuracy;
3. Quality control program requirements; and
4. Time requirements for maintaining laboratory analysis records and quality control data.

The DNR is directed to issue an initial certification to a laboratory for a specified test category if the laboratory meets the standards and criteria imposed by the DNR.

Certification of laboratories is renewed annually and guidelines are provided for suspension and revocation of laboratory certification.

E. REGISTRATION PROCEDURES

Registration procedures are provided for:

1. Laboratories which perform tests on their own behalf or on the behalf of a subsidiary or other corporation under common ownership or control; or
2. Laboratories owned or controlled by a municipality or two or more municipalities which perform tests solely on behalf of the municipality or municipalities.

In general, the criteria for laboratory registration are less stringent than those for laboratory certification and compliance with the criteria may not always be required. A laboratory may be registered by the DNR upon application and a demonstration that the laboratory has complied, where required, with criteria established by DNR for the following:

1. An accepted methodology, if tests are conducted in connection with a covered program;
2. Reference samples of tests, if required;
3. Self-audits and quality control programs; and
4. Time requirements for maintaining laboratory analysis records and quality control data.

Registration of laboratories is renewed annually and procedures are provided for suspension or revocation of registration. A laboratory which is either eligible or required to be registered may elect to apply for DNR laboratory certification.

F. FUNDING

Act 410 authorizes the DNR to promulgate by rule a graduated schedule of fees for certified and registered laboratories to recover the costs of administering the program. Act 410 also appropriates $17,500 to the DNR to establish the program and to administer the program until fees can be collected. This appropriation must be repaid from fees which are collected.

Part VI
Miscellaneous Provisions

This Part of the Memorandum contains a description of those parts of Act 410 relating to miscellaneous aspects of groundwater management.

A. MUNICIPALITIES

Zoning: Cities, villages, towns and counties are authorized, as one of the purposes of zoning, "to encourage the protection of groundwater resources" [ss. 59.97 (1), 60.74 (1) (a) 7 and 62.23 (7) (c)].

Real Property Values: Act 410 requires an assessor, when determining the market value of real property with a contaminated well or water supply, to consider the time and expense necessary for repairing or replacing the well or water supply when calculating the diminution of the market value of real property attributable to the contamination [s. 70.327].

In addition, the Act requires that an assessor consider the "environmental impairment" of the value of a property because of the presence of a solid or hazardous waste disposal facility [s. 70.32 (1m)].

Septage Disposal Ordinances: Counties are authorized, subject to supervision by the DNR, to regulate the disposal of septage (material pumped from septic tanks and holding tanks) on land. The county ordinance must include criteria and disposal procedures which are identical to DNR rules in ch. NR 113, Wis. Adm. Code. A city, village or town may only adopt a septage disposal ordinance if the county does not do so. The city, village or town ordinance is subject to the same restrictions as the county ordinance [s. 146.20 (5m)].

Regulation of Well Construction: Counties are authorized, subject to supervision by the DNR, to regulate well construction and pump installation for certain private wells. Cities, villages and towns are specifically denied authority to adopt regulations on this subject, regardless of whether the county adopts an ordinance [ss. 59.067 (5), 59.07 (51) and 162.07 and SECTIONS 2038 (6) and 2204 (38) (b)].

Septage Disposal in Municipal Sewage Systems: Municipal sewage systems are required, under certain circumstances and within certain limits, to accept septage from licensed septage disposers to minimize the land disposal of septage during months when the ground is frozen [s. 144.08 and SECTION 2204 (38) (a)].

Landfill Official Liability: Governmental officers and employees who are engaged in the planning, management, operation or approval of solid or hazardous waste disposal facilities are exempt from civil liability for good faith actions taken within the scope of their official duties [s. 144.795].

B. DEPARTMENT OF TRANSPORTATION

The DOT is required to regulate the bulk storage of salt and other chlorides, and mixtures of sand and salt, intended for application to highways during winter months [s. 85.16 and SECTION 2051 (1m)].

C. DEPARTMENT OF INDUSTRY, LABOR AND HUMAN RELATIONS

Storage of Flammable and Combustible Liquids: The DILHR is required to regulate the storage of flammable and combustible liquids for the purposes of groundwater and surface water protection [s. 101.09 and SECTION 2025 (2m)].

Plumbing Regulations: The DILHR is required to consider and adopt, as part of its regulation of plumbing, specific protections for groundwater and surface water [s. 145.13].

Unused, Underground Petroleum Product Storage Tanks: The DILHR is required to compile an inventory of unused, underground petroleum product storage tanks. An appropriation of $25,000 and 1.0 project position are provided for this purpose [ss. 20.445 (1) (dm) and 101.142 and SECTION 2025 (3)].

D. DEPARTMENT OF AGRICULTURE, TRADE AND CONSUMER PROTECTION

Storage of Fertilizer and Pesticides: The DATCP is required to regulate the storage of bulk quantities of fertilizer and pesticides for the purposes of groundwater and surface water protection [s. 94.645 and SECTION 2002 (2m)].

Bottled Drinking Water: Bottled drinking water which is manufactured or bottled for sale or distribution in this state must comply with state drinking water standards and health-related groundwater enforcement standards which are adopted by the DNR. The DATCP is responsible for administering and enforcing these requirements and may require the testing of bottled drinking water for substances which have contaminated water supplies or groundwater in the vicinity of the source of the water supply for the bottled drinking water [s. 97.34].

Pesticide Manufacturer License Fee Surcharge: The primary producer of a pesticide which is sold or distributed in Wisconsin must pay an annual license fee surcharge of $2,000 in addition to the basic $100 license fee. Approximately 40 primary producers will pay this fee. The revenues from this fee are appropriated to the DATCP for pesticide regulation [s. 94.681 (5)].

E. DEPARTMENT OF NATURAL RESOURCES

Recycling Research Projects: The DNR is authorized to waive compliance with feasibility report, plan of operation, licensing, long-term care and negotiation-arbitration requirements for projects which are intended for research to evaluate recycling and reuse of certain high-volume industrial waste [s. 144.44 (1m) (c), (2) (b) (intro.) and (7)].

Cooperative Agreements with American Indian Tribes: The DNR may negotiate and enter into cooperative agreements with American Indian tribes and bands for the purpose of establishing groundwater monitoring programs and groundwater regulatory programs on the lands of any American Indian tribe or band [s. 160.36].

Scenic Urban Waterways: Act 410 creates a program and appropriates $200,000 for scenic urban waterway designation and management within the DNR. The Act includes the designation of the Illinois Fox River and its watershed as a scenic urban waterway [ss. 20.370 (2) (ka) and 30.275].

Septage Disposal: Several changes are made in the state septage disposal regulations, which are administered by the DNR. The changes include new licensing requirements and fees [s. 146.20].

F. GROUNDWATER COORDINATING COUNCIL

Several state agencies are involved in nonregulatory activities relating to groundwater, such as education, laboratory analysis, data management and research. Act 410 creates a groundwater coordinating council to coordinate these agency activities.

The council is an independent body, attached to the DNR for administrative purposes only. The council has eight members, representing the Governor and those state agencies with responsibilities relating to groundwater. The members of the council, in addition to the Governor's representative, are the Secretaries of DNR, DILHR, DATCP, DOT, DHSS and the President of the University of Wisconsin System and the State Geologist, or their designees.

The main purpose of the groundwater coordinating council is to advise and assist state agencies in the coordination of nonregulatory programs and the exchange of information relating to groundwater. The council is required to present an annual "state of the groundwater" report to the members of the "appropriate" standing committees of the Legislature and the Governor and to the head of each agency with membership on the council. In addition, the council is required to review the provisions of Act 410 and report to the Legislature on implementation of the Act by January 1, 1989 [ss. 15.347 (13) and 160.50 and SECTION 2038 (9)].

G. PESTICIDE REVIEW BOARD

Rule Review: In general, the approval of the Pesticide Review Board (comprised of the Secretaries, or their designees, of the DATCP, DHSS and DNR) is required for rules regulating pesticides. Any rules regulating pesticides which are developed under ch. 160, Stats., are exempt from review and approval by the Pesticide Review Board [ss. 29.29 (4), 94.69 (10) and 140.77 (1)].

H. UNIVERSITY OF WISCONSIN SYSTEM

Fertilizer Research Funds: Existing law is amended to clearly allow fertilizer research funds to be used by the University of Wisconsin System for research on groundwater problems which may be related to fertilizer use [s. 94.64 (8m) (a)].

I. PUBLIC INTERVENOR

Authority: The Public Intervenor is given authority to initiate actions and proceedings before any agency or court. However, Act 410 specifies that the Public Intervenor does not have the authority to initiate any action or proceeding concerning the issuance of obligations by the Building Commission [ss. 18.13 (4) and 165.075].

Advisory Committee: The Public Intervenor Advisory Committee is established by statute [s. 165.076].

J. FOX RIVER MANAGEMENT COMMISSION

Act 410 creates a Fox River management commission and provides the commission with the authority to: (1) negotiate and enter into agreements with the federal government; and (2) manage the Fox River locks and facilities [ss. 15.01 (4), 15.06 (1) (e) and (3) (a) 4, 15.341 (5), 15.345 (5), 20.370 (1) (da) and (dj) and 30.93 and SECTION 2038 (7)].

Appendix 1
History of the Special Committee on Groundwater Management

The Legislative Council established the Special Committee on Groundwater Management on January 28, 1982. The Special Committee was established by the Legislative Council in response to a request to conduct the study in SEC. 121 of Ch. 374, Laws of 1981 (an act relating to solid and hazardous waste management), and in a letter from Representative Mary Lou Munts. The Committee consisted of five Representatives, three Senators and eight public members as follows:

Representatives

Mary Lou Munts, Madison
  Chairperson
Patricia Goodrich, Berlin
  Secretary
Thomas Crawford, Milwaukee
Gervase Hephner, Chilton
Mary Panzer, West Bend

Senators

Joseph Strohl, Racine
  Vice-Chairperson
David Helbach, Stevens Point
David Opitz, Port Washington

Public Members

Roger Cliff, Madison	Terry Kakida, Milwaukee
James Derouin, Madison	Lonnie Kragwold, Amherst
Myron Ehrnardt, Oakfield	Tom Kunes, Madison
Professor James Hoffman, Oshkosh	Douglas Mormann, Stevens Point

The Special Committee was given a broad directive by the Legislative Council to make recommendations on the state's groundwater management policies. The Special Committee was directed to study and submit recommendations on the following issues:

1. Formulating a statutory goal statement clarifying the state's groundwater management policies;
2. Directing state agencies to develop groundwater management rules, where applicable, in compliance with the statutory groundwater policies;
3. Formulating the means to implement groundwater management procedures;
4. Establishing statutory authority and regulatory programs to resolve groundwater management problems; and
5. Suggesting programs which are not solely regulatory, such as public education programs or coordinating councils.

The Legislative Council also established a Technical Advisory Committee (TAC) to assist the Special Committee in completing its assignment. The Legislative Council appointed representatives of the following seven state agencies to serve on the TAC:

Dr. Henry Anderson
DHSS
Linda Bochert
DNR
Professor Stephen Born
University of Wisconsin—Extension
Thomas Dawson
DOJ
Orlo R. Ehart
DATCP
Dave Fredrickson
DILHR
Meredith Ostrom
Wisconsin Geological and Natural History Survey

Chairperson Munts established a Subcommittee on Compensation for Groundwater Damages to review existing methods whereby individuals can obtain compensation for groundwater damages and to prepare recommendations for changes in existing compensation methods or for the creation of new compensation programs. The Subcommittee consisted of the following members:

Representatives

Thomas Crawford
Milwaukee

Mary Lou Munts
Madison

Public Members

Roger Cliff Madison	James Derouin Madison

TAC Members

Dr. Henry Anderson DHSS	Thomas Dawson DOJ
Linda Bochert DNR	Orlo R. Ehart DATCP

The Special Committee held nine meetings at the State Capitol and one meeting in Stevens Point which included an afternoon tour of areas affected by groundwater contamination and an evening public hearing. The TAC held 17 meetings in Madison and the Subcommittee held 15 meetings in Madison.

As a result of these extensive deliberations, the final recommendations of the Special Committee were prepared in bill draft form and approved by the Special Committee at its May 16, 1983 meeting. At its June 29, 1983 meeting, the Legislative Council voted to introduce the Special Committee's proposal in the 1983 Legislative Session. The bill was introduced on July 28, 1983, as 1983 Assembly Bill 595.

Appendix 2
Legislative Action on 1983 Assembly Bill 595

Assembly Bill 595

Assembly Bill 595, introduced by the Legislative Council, was referred to the Assembly Committee on Environmental Resources. The Bill contained the following provisions:

1. A state-funded program to provide compensation for any person whose well is rendered unusable as the result of groundwater contamination.
2. A framework for the development and implementation of groundwater protection standards for substances detected in, or with the potential to enter, the groundwater resources of the state.
3. Several minor substantive provisions concerning groundwater regulation.

Assembly Substitute Amendment 1

The Assembly Committee on Environmental Resources held two public hearings on the Bill and, on February 24, 1983, the Committee introduced Assembly Substitute Amendment 1 and voted to adopt it by a vote of Ayes, 9; Noes, 2; and Absent, 0. The Committee then recommended passage of Assembly Substitute Amendment 1 to Assembly Bill 595 by the same vote. Assembly Substitute Amendment 1 to Assembly Bill 595 contained the following provisions, in addition to the original provisions of Assembly Bill 595:

1. A requirement to purchase insurance for the purpose of replacing contaminated wells. This proposal was substituted for the state-funded compensation program;
2. An environmental repair fund and program to conduct inventories and provide for the cleanup of contaminated waste disposal sites in the state;
3. A program to register or certify laboratories which conduct environmental sampling and analyses;
4. A groundwater monitoring program; and
5. Fees to support a groundwater fund which is used to establish and implement groundwater standards, to implement the groundwater monitoring program and to implement the environmental repair fund program.

Assembly Floor Amendments

The Assembly adopted Assembly Substitute Amendment 1 to Assembly Bill 595 and 27 of the 89 amendments which were introduced. The 27 amendments to Assembly Substitute Amendment 1 incorporated the following major provisions in the Bill:

1. Statutory authority for the DATCP to administer and enforce an animal waste management program and restrictions on the authority of the DNR to adopt animal waste regulations;
2. A state-funded program to provide compensation for contaminated wells. This proposal was substituted for the requirement to obtain well contamination insurance; and
3. An optional program which permits counties to regulate the installation of private wells.

Senate Substitute Amendment 1

The Senate Committee on Energy and Environmental Resources adopted and recommended for passage, by a vote of Ayes, 5; Noes, 2; and Absent, 0, Senate Substitute Amendment 1 to Assembly Bill 595. This Substitute Amendment made the following changes to Engrossed 1983 Assembly Bill 595:

1. The authority for DATCP to regulate animal waste was deleted and the authority for such regulation was restored to the DNR; and
2. A program requiring the DILHR to inventory unused, underground petroleum product storage tanks was created.

Senate Floor Amendments

The Senate adopted Senate Substitute Amendment 1 to Assembly Bill 595 and 10 of the 19 amendments which were introduced. The major changes made by the 10 amendments to Senate Substitute Amendment 1 were as follows:

1. The DATCP was again granted authority to regulate animal waste and the DNR authority for such regulation was restricted; and
2. A program requiring the DATCP to set standards for the testing of bottled drinking water was created.

Assembly Concurrence

The Assembly concurred in Engrossed Senate Substitute Amendment 1 to Assembly Bill 595 without further amendment.

Governor's Signature and Vetoes

The Bill was signed by the Governor on May 4, 1984. The act was numbered 1983 Wisconsin Act 410 and took effect on May 11, 1984.

In signing the Bill, the Governor exercised his item veto authority several times. The principal item veto was the deletion of the transfer of animal waste regulations from the DNR to the DATCP. In the Veto Review Session on May 24, 1984, the Legislature failed to override any of the Governor's vetoes.

APPENDIX E
Suffolk County Sanitary Code Article 7–Water Pollution Control

---

Section 701. Declaration of Policy

The designated best use of all groundwaters of Suffolk County is for public and private water supply, and of most surface waters for food production, bathing and recreation. The federal government has officially designated the aquifer below Suffolk County as a sole-source for water supply. Therefore, it is hereby declared to be the policy of the County of Suffolk to maintain its water resources as near to their natural condition of purity as reasonably possible for the safeguarding of the public health and, to that end, to require the use of all available practical methods of preventing and controlling water pollution from sewage, industrial and other wastes, toxic or hazardous materials, and stormwater runoff.

Section 702. Statement of Purpose

It is the intent and purpose of this article to safeguard all the water resources of the County of Suffolk, especially in deep recharge areas and water supply sensitive areas, from discharges of sewage, industrial and other wastes, toxic or hazardous materials and stormwater runoff by preventing and controlling such sources in existence when this article is enacted and also by preventing further pollution from new sources under a program which is consistent with the above-stated Declaration of Policy.

Section 703. Definitions

Whenever used in this article, unless otherwise expressly stated, or unless the context or subject matter requires a different meaning, the following terms shall have the respective meanings set forth or indicated.

A. **Board** means the Suffolk County Board of Health.

B. **Commissioner** means the Commissioner of the Suffolk County Department of Health Services.

C. **Communal Sewage System** means a series of sanitary intercepting sewers or intercepting collecting sewers, pumping stations, sewage treatment plants, and associated pollution control facilities for the conveyance, treatment, and disposal of sewage operated by a person other than a municipality.

D. **Deep Recharge Area** means a geographic area of Suffolk County that contributes recharge water to a deep groundwater flow system, thus replenishing the quantity and affecting the quality of the long-term water supply. These areas are identified as Groundwater Management Zones I, II, III and V.

E. **Department** means the Suffolk County Department of Health Services.

F. **Discharge** means to release by any means or to relinquish control in a manner that could result in a release to the surface waters, groundwaters, surface of the ground, or below ground.

G. **Disposal System** means any plumbing or conveyances which result in or are capable of resulting in a discharge of sewage, industrial wastes, toxic or hazardous materials, stormwater runoff, cooling water or other wastes. This includes but is not limited to septic tanks, leaching pools, sumps, tile fields, holding tanks, outfalls and connecting piping.

H. **Groundwater Management Zone** means any of the areas delineated in Suffolk County by the "Long Island Comprehensive Waste Treatment Management Plan (L.I. 208 Study)," as revised by the "Long Island Groundwater Management Plan," and subsequent revisions adopted by the Board identifying differences in regional hydrogeologic and groundwater quality conditions. The boundaries of the Groundwater Management Zone are set forth on a map adopted by the Board filed in the Office of the Commissioner in Hauppauge, New York.

I. **Housebarge** means the same as Houseboat except that a housebarge has no self-contained mechanical method of propulsion.

J. **Houseboat** means a floating structure used as a dwelling with a self-contained mechanical method of propulsion, not primarily designed to be a means of locomotion over water. The design criteria shall be generally accepted standards of naval architecture.

K. **Industrial Waste** means any liquid, gaseous, or solid waste substance or a combination thereof resulting from any operation or process of industry, manufacturing, trade or business or from the development or recovery of any natural resources, which may cause or might reasonably be expected to cause pollution of the water resources of the County of Suffolk in contravention of the requirements of this article.

L. **Municipal Sewage System** means the series of sanitary intercepting sewers or intercepting collecting sewers, pumping stations, sewage treatment plants, or pollution control facilities, drains and other facilities, connections and equipment or any combination of the aforementioned, for the conveyance, treatment and disposal of sewage operated by the County of Suffolk or a municipality within the County of Suffolk.

M. Offensive Material means any sewage or non-sewage fecal matter, urine, garbage, waste, or any putrescible organic matter, scavenger waste, the contents of private or individual sewage disposal systems, either liquid or solid, or other substances or liquid which may adversely affect health.

N. Other Wastes means refuse, spillage and the leaching from these materials, oil, tar, acids, chemicals, and all other discarded matter which may reasonably be expected to cause pollution of the waters of the County of Suffolk.

O. Private or Individual Sewage Disposal System means a water-flush facility for the disposal of sewage which does not connect either with a municipal or communal sewage system. This includes, but is not limited to, septic tanks, leaching pools and tile fields.

P. Restricted Toxic or Hazardous Materials shall mean the following toxic or hazardous chemicals that have been or could be expected to be detected in the groundwater, or in discharges to the groundwater, of Suffolk County. This definition applies to these substances alone or in combination, solution or mixture with other substances, or chemically compounded with other elements or compounds.

Arsenic
Barium
Benzene
Bromobenzene
Bromodichloromethane
Bromoform
Cadmium
Carbon Tetrachloride
Chlorobenzene
Chlorodibromomethane
Chloroform
Chlorotoluene
Chromium
Cis 1,2 Dichloroethylene
Creosotes
Cyanide
Dichlorobenzene
1,1 Dichloroethane
1,2 Dichloroethane
1,3 Dichloroethane
1,1 Dichloroethylene
1,2 Dichloropropane
p-Diethylbenzene
Ethylbenzene
p-Ethyltoluene
Fluoride
Freon 113
Lead
Mercury
Methylene Chloride
Nickel
Pesticides
Petroleum Distillates
Phenols
Phthalates
Roadway Deicing Salt
Silver
Styrene
Tetrachloroethylene
1,2,4,5 Tetramethylbenzene
Toluene
1,2,3 Trichlorobenzene
1,2,4 Trichlorobenzene
1,1,1 Trichloroethane
1,1,2 Trichloroethane
1, 1, 2 Trichloroethylene
1, 2, 3 Trichloropropane
1, 2, 4 Trimethylbenzene
1, 3, 5 Trimethylbenzene
Vinyl Chloride
Xylenes

All other halogenated hydrocarbon compounds.

Q. **Sewage** means the water-carried human or animal wastes from residences, buildings, industrial establishments or other places, together with such groundwater infiltration and surface water as may be present. A mixture of sewage as herein defined and industrial wastes or other wastes as defined above may be considered industrial wastes or commingling within the meaning of this article.

R. **Stormwater Runoff** means the portion of total precipitation that travels over natural and developed land surfaces (e.g., woodlands, lawns, farms, gardens, roofs, driveways, parking lots, roads, etc.) transporting contaminants that may be present.

S. **Temporary Disposal System** means a system for the disposal of sewage where such system is intended for use for a specified period of time prior to completion of the construction of an approved sewage treatment and disposal system.

T. **Toxic or Hazardous Materials** shall mean the same as defined in Article 12 of this Code.

U. **Toxic or Hazardous Wastes** shall mean the same as defined in Article 12 of this Code.

V. **Treatment System** means a system designed to reduce or alter the contaminant content of sewage or industrial waste for the purpose of permitting the discharge of some portion of said waste.

W. **Water Supply Sensitive Areas** means:

1. A groundwater area separated from a larger regional groundwater system where salty groundwater may occur within the Upper Glacial aquifer, and where deepening of private wells and/or the development of community water supplies may be limited; or
2. Areas in close proximity to existing or identified future public water supply wellfields. In general, for the purposes of this article, "close proximity" shall mean within 1,500 feet upgradient or 500 feet downgradient of public supply wells screened in the Upper Glacial aquifer.
3. A limited water budget area, not underlined by fresh Magothy, defined by published reports acceptable to the commissioner.
4. The areas described in items W.1., 2., 3., above are set forth on a map adopted by the Board filed in the Office of the Commissioner in Hauppauge, New York.

Section 704. Powers of the Commissioner

The commissioner may:

A. make or cause to be made, any investigation which, in his opinion, is needed for the enforcement of this article or for controlling or reducing the potential for contamination of the waters of the county from sewage, industrial or other wastes, toxic or hazardous materials and/or stormwater runoff;

B. approve, with conditions, non-residential structures, processes, facilities and activities in deep recharge areas and water supply sensitive areas to assure compliance with Section 706. Such conditions shall be embodied in convenants running with the land as specified in the Department's standards;

C. promulgate the established standards and schedules to effect the purpose of this article;

D. order the posting of a performance bond or other undertaking either prior to or subsequent to the construction or operation of an industrial facility within Suffolk County on a case-by-case basis if evidence indicates such may be necessary to protect water resources from the adverse effects of operating such a facility.

Section 705. General Restrictions and Prohibitions

A. Construction of a Disposal System

1. It shall be unlawful for any person to construct, reconstruct, install or substantially modify any disposal system without first having obtained a permit therefor issued by or acceptable to the commissioner.

2. Section 705.A.1 does not apply to stormwater disposal systems unless there is an actual or potential discharge into the system of industrial wastes, toxic or hazardous materials, or sewage.

B. Discharge

1. It shall be unlawful for any person to discharge sewage, industrial wastes, offensive materials, toxic or hazardous materials or other wastes to any surface waters or groundwaters, to the surface of the ground or to a disposal system unless such discharge is specifically in accordance with a State Pollutant Discharge Elimination System (SPDES) Permit or other permit issued by or acceptable to the commissioner for that purpose.

2. No permits, as stipulated in Section 705.B.1, are required for the following types of discharges:

a. discharge of sewage from an existing residential structure to a private or individual sewage disposal system, or from any residential structure, houseboat or housebarge to a communal sewage system or municipal sewage system that does not contravene standards or result in a public health nuisance;

b. discharge of sewage from a commercial or industrial facility to a communal sewage system or municipal sewage system;

c. discharge of stormwater to a disposal system unless there is an actual or potential discharge into the system of industrial wastes or toxic or hazardous materials or sewage.

3. For existing discharges not prohibited by law prior to the effective date of this article, a permit shall be obtained within the time limits provided in Section 707.

C. Construction or Operation of a Treatment System

1. It shall be unlawful for any person to construct, modify or operate a treatment system without first obtaining a permit therefor issued by or acceptable to the commissioner.

D. Commingling

1. It shall be unlawful for any person to commingle stormwater runoff, cooling water, sewage or industrial wastes in any disposal system not approved for that purpose pursuant to this article.

E. Stormwater Discharges

1. It shall be unlawful for any person to develop or use land in such a manner as to cause stormwater runoff from that land to become contaminated and discharged in contravention of the other provisions of this article.

Section 706. Deep Recharge Areas and Water Supply Sensitive Areas

The following additional restrictions and prohibitions shall apply in deep recharge areas and water supply sensitive areas.

A. It shall be unlawful for any person to discharge any restricted toxic or hazardous materials or to discharge industrial wastes from processes containing restricted toxic or hazardous materials to the groundwaters, to the surface of the ground, beneath the surface of the ground, to a municipal or communal sewage system, or to a disposal system except as follows:

1. application of fertilizers, pesticides or other agricultural chemicals approved for that purpose by the appropriate state and federal agencies; or
2. application of road surfacing or road construction materials or deicing salts to roadways, walkways, and parking areas; or
3. discharge from an establishment to a municipal or communal sewage system with effluent disposal to marine surface waters or recharge outside of the deep recharge areas and water supply sensitive areas, and the following minimum requirements are satisfied pursuant to a permit issued by or acceptable to the commissioner:
   a. Dual plumbing systems shall be installed, one for the sanitary wastes and one for industrial wastes.
   b. Sampling access approved by the administrative head of the municipal or communal sewage system and the Department shall be provided for both the sanitary and industrial waste systems.
   c. The administrative head of the municipal or communal sewage system, with approval of the Department, shall determine which industrial wastes are acceptable to "hold and haul" and which require pretreatment prior to discharge to the collection system in order to assure compliance with the applicable sewer use ordinance.

d. Personnel authorized by the administrative head of the municipal or communal sewage system or other individual(s) acceptable to the commissioner, shall operate at each establishment its pretreatment facility for industrial wastes prior to discharge to the collection system.

e. Only batch pretreatment of industrial wastes will be permitted. Batch facilities and facilities for storage of drums containing toxic or hazardous wastes shall be located in an area accessible at all times by district personnel, in or adjacent to the industrial building, with heat and power provided by the owner.

f. Personnel authorized by the administrative head of the municipal or communal sewage system or other individual(s) acceptable to the commissioner, will be responsible for collection and disposal of pretreatment sludges, and other "hold and haul" materials.

g. The owner shall allow the personnel authorized by the municipal or communal sewage system or other individual(s) acceptable to the commissioner, access, from time to time, to wet process areas to perform their duties and inspections.

h. Industrial process-area floors shall be provided with adequate means to contain any spill of restricted toxic or hazardous materials. The design of containment facilities shall be subject to the approval of the commissioner.

i. A minimum of four (4) groundwater monitoring wells shall be installed at the owner's expense.

j. Financial assurance shall be provided to pay for cleanup of spills. This cost shall be entered as a judgment upon notice against the owner, occupant, tenant, or lessee responsible for such spill or spills.

B. It shall be unlawful to use or store any restricted toxic or hazardous materials on any premises except as follows:

1.a. the intended use of the product stored is solely for on-site heating, or intermittent stationary power production such as stand-by electricity generation or irrigation pump power; and

1.b. the facility for such storage is intended solely for the storage of kerosene, number 2 fuel oil, number 4 fuel oil, number 6 fuel oil, diesel oil or lubricating oil; and

1.c. the facility for such storage is constructed in accordance with the construction standards of Article 12 of the Suffolk County Sanitary Code for non-petroleum hazardous materials; and

1.d. the materials so stored are not industrial wastes from processes containing restricted toxic or hazardous materials; and

1.e. the materials stored are not intended for resale; or

2.a. the materials so stored are in containers where the total liquid capacity stored at any time does not exceed 250 gallons and where the dry storage in bags, bulk or small containers does not exceed 2,000 pounds; or

3.a. the materials so stored are intended solely for treatment or disinfection of water or sewage in treatment processes located at the site; or

4.a. the materials are stored solely incident to retail sales on premises and are not processed, pumped, packaged, or repackaged at the site; or

5.a. the materials are stored at a service station or similar installation solely incident to the distribution of gasoline, kerosene, diesel oil or other petroleum products for motor vehicular uses and repair; and

5.b. the facility for such storage is constructed in accordance with construction and monitoring standards of Article 12 of the Suffolk County Sanitary Code for non-petroleum hazardous materials; or

6.a. the materials are stored at an establishment for which a permit has been secured in accordance with Section 706.A.3, and a permit for such storage has been granted by the Department;

7.a. the materials are stored on a farm site solely incident to on-premises use, and consist of fertilizers, pesticides, or other agricultural chemicals to be applied in accordance with the provisions of Section 706.A.1.

C. The provisions of Sections 706.A and 706.B of this article shall be applicable:

1. immediately for all non-residential facilities which have not been approved, constructed, or put into operation prior to the effective date of this article; and
2. immediately for all non-residential facilities which were approved, constructed, or put into operation prior to the effective date of this article upon:

a. any change in use or process which results in an increase of mass loading in the discharge or restricted toxic or hazardous materials, or introduces a toxic or hazardous material not previously discharged; or

b. any change in use or process which results in an increase of the storage or change of type of restricted toxic or hazardous materials.

D. When upgraded in accordance with the time schedule specified in Article 12, existing facilities, including those for petroleum products, not otherwise covered by items 706.A, 706.B or 706.C, above, shall conform to the standards of Article 12 for non-petroleum hazardous materials. These requirements do not apply to facilities upgraded in accordance with Article 12 prior to the effective date of this article.

Section 707. Permits

A. All permits required by this article shall be applied for in accordance with the provisions of Article 3 of the Suffolk County Sanitary Code.

B. All persons required to obtain a permit by reason of any law, rule or regulation in effect prior to the effective date of this article shall be governed by such law, rule or regulation in determining when said permit shall be obtained.

C. All persons newly required to obtain a permit by this article due to any act or condition in existence as of the date this article becomes effective, shall apply for said permit within one (1) year of that date.

D. All persons required to obtain a permit by this article due to any act or condition not in existence on the effective date of this article must apply for and receive said permit prior to undertaking such act or creating such condition.

Draft 8/14/84
Approved for Publication 8/29/84
1/22/85
2/27/85
4/24/85
Adopted 5/22/85

APPENDIX **F**

Amendment to Environmental Conservation Law in Relation to Land Burial and Disposal of Solid Wastes in Nassau and Suffolk Counties, New York

---

1983–1984 Regular Sessions

IN ASSEMBLY

March 28, 1983

Introduced by M. of A. NEWBURGER, FINK, HINCHEY, YEVOLI, HALPIN—Multi-Sponsored by—M. of A. BIANCHI, BRANCA, ENGEL, FELDMAN, FLANAGAN, GRANNIS, HARENBERG, HEVESI, HOCHBRUECKNER, JACOBS, KOPPELL, MURTAUGH, ORAZIO, PATTON, PILLITTERE, RETTALIATA, SAWICKI, SEMINERIO, SIEGEL—(at request of the Governor)—read once and referred to the Committee on Environmental Conservation—amended on the special order of third reading, ordered reprinted as amended, retaining its place on the special order of third reading

AN ACT to amend the environmental conservation law, in relation to land burial and disposal of solid waste in the counties of Nassau and Suffolk

---

EXPLANATION—Matter in italics (underscored) is new; matter in brackets [ ] is old law to be omitted.

The People of the State of New York, represented in Senate and Assembly, do enact as follows:

Section 1. Legislative findings. The legislature hereby finds that the land burial and disposal of domestic, municipal and industrial solid waste poses a significant threat to the quality of groundwater and therefore the quality of drinking water in the counties of Nassau and Suffolk. This threat is particularly dangerous since the potable water supply for the counties is derived from a sole source aquifer. Scientific evidence and analysis have identified the incapacity of land burial and disposal to isolate leaching chemicals and gases from the surrounding environment over the long term. Resource recovery of these wastes poses minimal threats to groundwater quality.

§2. The environmental conservation law is amended by adding a new section 27-0704 to read as follows:

§27-0704. Land burial and disposal in the counties of Nassau and Suffolk; special provisions.

1. Definitions. As used in this section the following terms shall have the following meanings:

a. "Clean fill" shall mean material consisting of concrete, steel, wood, sand, dirt, soil, glass, or other inert material designated by the commissioner.

b. A "deep flow recharge area" shall mean a sensitive recharge area within the counties of Nassau and Suffolk within the boundaries of hydrogeologic zones I, II and III as defined in the Long Island Comprehensive Waste Treatment Management Plan of nineteen hundred seventy-eight.

c. "Downtime waste" shall mean any treatable or burnable waste accumulated during a scheduled or unscheduled maintenance period of a treatment facility.

d. "Hazardous waste" shall be defined as promulgated by the provisions of section 27-0903 of this article.

e. "Landfill" shall mean a disposal facility at which solid waste, or its residue after treatment, is intentionally placed and at which, waste shall remain after closure.

f. "Long Island Comprehensive Waste Treatment Management Plan of nineteen hundred seventy-eight" shall mean the study prepared by the Long Island Regional Planning Board pursuant to section two hundred eight of the federal water pollution control act.

g. "Treatment facility" shall mean resource recovery, incineration, composting, or other process as approved by the commissioner through which solid waste is put in order to reduce volume and toxicity.

h. "Untreatable waste" shall mean that material that because of its size or composition cannot be processed by a treatment facility.

2. The Long Island Comprehensive Waste Treatment Management Plan of nineteen hundred seventy-eight shall be kept on file in the office of the commissioner. The hydrogeologic zones and their attendant boundaries as specified in

the aforementioned plan are hereby adopted. Any changes made in the boundaries and accepted by the commissioner shall be considered as automatically adopted for the purposes of this section.

3. On or after the effective date of this section and except as provided herein, no person shall commence operation, including site preparation, of a new landfill or of an expansion to an existing landfill which is located in a deep flow recharge area. However, the commissioner, after conducting a public hearing, may approve a limited expansion of any existing landfill in a deep flow recharge area for the sole purpose of providing for solid waste disposal capacity prior to the implementation of a resource recovery system. The commissioner shall not approve any such expansion unless he finds that the owner of such landfill is a municipality that is implementing a resource recovery system which is acceptable to the commissioner and which will be operational no later than seven years after the effective date of this section and that no other feasible means of solid waste management is available, taking into account technological, economic and other essential factors.

4. On or after the effective date of this section, no person shall commence operation, including site preparation, of a new landfill or of an expansion to an existing landfill, which is located in the county of Nassau or Suffolk outside of deep flow recharge areas unless:

a. The commissioner has made an affirmative determination that such landfill will not pose a threat to groundwater quality; and

b. The owner or operator of the landfill has posted a financial guarantee such as, but not limited to, pollution liability insurance, sureties, performance bonds and/or trust funds acceptable to the commissioner securing the cost of corrective treatment, or the development of alternative water sources, should such landfill become a source of groundwater, surface water, or air pollution. The size of the financial guarantee, the financial stability of the surety, and the terms of posting shall be determined by the commissioner. Financial surety shall also be arranged to ensure the proper operation and maintenance of leachate and other collection and treatment systems for a period of time, as determined by the commissioner, after a landfill is closed; and

c. The landfill is underlain by two or more natural and/or synthetic liners each with provisions for leachate collection, and has a treatment and disposal system, all of which are approved by the commissioner. Any natural clay liners shall have a maximum compacted thickness of two feet and all liners shall have a maximum hydraulic conductivity not to exceed one times ten to the minus seven centimeters per second. If the landfill uses two synthetic liners, the department shall require that the liners are of different chemical compositions; and

d. The landfill is designed and operated to minimize the migration of methane gas or other gases beyond the facility boundaries so as to avert the creation of a nuisance or a danger to property or public health; and

e. The landfill is prohibited from accepting industrial commercial or institutional solid or liquid waste that is hazardous; and

f. The landfill is not located in a freshwater wetland, tidal wetland or floodplain as identified by the department.

g. Except as provided herein, the landfill accepts only material which is the product of resource recovery, incineration or composting. Downtime waste and wastes that are untreatable by a resource recovery system may be disposed of when handled as provided in this paragraph. Downtime waste and untreatable waste that is landfilled may only be deposited in a special disposal area that is located and constructed so as to segregate these wastes and minimize their effect on residents of the surrounding area. Not more than ten percent of the annual rated capacity of a resource recovery facility may be disposed of as downtime waste per year. However, up to ten percent of the annual rated capacity of more than one resource recovery facility may be so disposed of at a single landfill.

Any such landfill may also accept wastes other than those authorized in this subdivision whenever such disposal is approved by the commissioner based upon a finding made after the opportunity for a public hearing that (i) no resource recovery facility is available to accept such waste; (ii) the owner of the landfill is making all reasonable efforts to implement a resource recovery system acceptable to the commissioner; and (iii) that the landfilling of such wastes will not have significant adverse environmental impacts. In granting any such approval, the commissioner shall impose conditions necessary to mitigate any adverse environmental impacts to the maximum extent practicable and shall impose a schedule under which the municipality shall implement an acceptable resource recovery system.

5. Within seven years of the effective date of this section, no person shall operate a landfill existing on the effective date of this section in the counties of Nassau and Suffolk unless:

a. The owner or operator of the landfill has posted a financial guarantee such as, but not limited to, pollution liability insurance, sureties, performance bonds and/or trust funds acceptable to the commissioner securing the cost of corrective treatment, or the development of alternative water sources, should such landfill become a source of groundwater, surface water or air pollution. The size of the financial guarantee, the financial stability of the surety, and the terms of posting shall be determined by the commissioner. Financial surety shall also be arranged to ensure the proper operation and maintenance of leachate and other collection and treatment systems for a period of time, as determined by the commissioner, after a landfill is closed; and

b. The landfill is underlain by two or more natural and/or synthetic liners each with provisions for leachate collection, and has a treatment and disposal system, all of which are approved by the commissioner. Any natural clay liners shall have a minimum compacted thickness of two feet and all liners shall have a maximum hydraulic conductivity not to exceed one times ten to the minus seven centimeters per second. If the landfill uses two synthetic liners, the department shall require that the liners are of different chemical composition; and

c. The landfill is designed and operated to minimize the migration of methane gas or other gases beyond the facility boundaries so as to avert the creation of a nuisance or a danger to property or public health; and

d. The landfill does not accept industrial, commercial or institutional solid or liquid waste that is hazardous; and

e. The landfill is not located in a freshwater wetland, tidal wetland or floodplain as identified by the department.

f. Except as provided herein, the landfill accepts only material which is the product of resource recovery, incineration or composting. Downtime waste and wastes that are untreatable by a resource recovery system may be disposed of when handled as provided in this paragraph. Downtime waste and untreatable waste that is landfilled may only be deposited in a special disposal area that is located and constructed so as to segregate these wastes and minimize their effect on residents of the surrounding area. Not more than ten percent of the annual rated capacity of a resource recovery facility may be disposed of as downtime waste per year. However, up to ten percent of the annual rated capacity of more than one resource recovery facility may be so disposed of at a single landfill.

If the landfill is located outside of the deep flow recharge area, such landfill may also accept wastes other than those authorized in this subdivision whenever such disposal is approved by the commissioner based upon a finding made after the opportunity for a public hearing that (i) no resource recovery facility is available to accept such waste; (ii) the owner of the landfill is making all reasonable efforts to implement a resource recovery system acceptable to the commissioner; and (iii) that the landfilling of such wastes will not have significant adverse environmental impacts. In granting any such approval, the commissioner shall impose conditions necessary to mitigate any adverse environmental impacts to the maximum extent practicable and shall impose a schedule under which the municipality shall implement an acceptable resource recovery system.

6. Notwithstanding the other provisions of this section, the commissioner may allow, by permit, the disposal of clean fill material in the counties of Nassau and Suffolk. Such material shall not be contaminated with hazardous wastes.

§3. This act shall take effect on the one hundred eightieth day next succeeding the date on which it shall have become a law.

APPENDIX G
About the Environmental Cleanup Responsibility Act (ECRA), New Jersey

The ECRA program is analogous to the home buyer protection programs instituted by many municipalities throughout the state. In an effort to maintain the quality of housing, many boroughs and townships instituted programs that require municipal inspections of those homes that are for sale. The inspection is used to determine flaws in the heating, plumbing, or electrical systems plus any other major hazard that may be existing. If the home owner does not fix the identified hazards, the municipality may withhold a certificate of occupancy to the buyer, thereby voiding the pending sale of the property. These local programs have been implemented in an effort to improve the quality of life. ECRA is no different.

The need for the ECRA program is well-documented. Too often, unsuspecting buyers of industrial establishments find out, after the purchase, that the facility has hidden environmental problems for which they, as the new owners, may not be responsible. Abandoned industrial establishments are often discovered to be the source of soil, ground water, and surface water contamination. For example, old landfills, unreported spills of hazardous substances, or unlined waste lagoons may have been used and then covered over prior to the sale. Unfortunately, some industrial establishments abandon operations, leaving state and federal agencies with the task of either taking remedial action on their own or instituting legal action to compel the company to clean up the site. Abandoned properties with serious contamination are often economically undevelopable and require state and federal tax dollars to pay for the decontamination of the site before private capital will be invested. In any case, it costs

SOURCE: New Jersey Department of Environmental Protection, Trenton, N.J.

New Jersey taxpayers millions of dollars. The bottom line of both the home buyer protection programs and the ECRA program is that those persons responsible for creating the problems must also be responsible for resolving them before the property is purchased by someone else. ECRA attempts to place that cleanup responsibility where it belongs by requiring industrial establishments to clean up their facilities as a precondition to closure, sale, or transfer of their operations. The costs of these cleanups will be borne now by the owner of the establishment and future problems at increased costs will be avoided.

ECRA is one of the most momentous environmental statutes in the country. It provides a healthy dose of preventive medicine, which will pay benefits to both the environment and the economy of New Jersey.

WHO IS REQUIRED TO COMPLY?

ECRA presents an initial two-part test to determine whether a particular business is an industrial establishment and thus subject to the act. First, a determination must be made that a particular place of business is engaged in operations involving the generation, manufacture, refining, transportation, treatment, storage, handling, or disposal of hazardous substances or hazardous wastes above or below ground. The regulations define hazardous substances as those elements and compounds, including petroleum products, defined as such by the department in Appendix A of N.J.A.C. 7:1E. Sewage and septage sludge are specifically exempted from the definition of hazardous substances contained in the act. Hazardous wastes are defined as those waste substances required to be reported to the department on the special waste manifest form pursuant to N.J.A.C. 7:26–7.4, designated as a hazardous waste pursuant to N.J.A.C. 7:26–8, or as otherwise provided by law. Both sets of regulations are included in this package.

The second part of the definition of industrial establishment states that the primary operations of the business being conducted on the site involved in the sale or closure must fall within a Standard Industrial Classification ("SIC") major group number of 22–39 inclusive, 46–49 inclusive, 51 or 76 as designated in the Standard Industrial Classification Manual prepared by the Office of Management and Budget in the Executive Office of the President of the United States. The SIC manual may be available at local libraries, or can be ordered from the Superintendent of Documents, U.S. Government Printing Office, Washington, D.C. 20402, Stock Number 4101-0066.

The act applies to the closure, termination, transfer, or sale of industrial establishments in the following SIC major group number categories:

SIC	Industry Category	SIC	Industry Category
22	Textile Mill Products	25	Furniture and Fixtures
23	Apparel	26	Paper and Allied Products
24	Lumber and Wood Products		

SIC	Industry Category	SIC	Industry Category
27	Printing, Publishing, and Allied Industries	37	Transportation Equipment
28	Chemicals and Allied Products	38	Measuring, Analyzing, and Controlling Instruments, Photographic Medical and Optical Goods
29	Petroleum Refining and Related Industries	39	Miscellaneous Manufacturing Industries
30	Rubber and Miscellaneous Plastics Products	46	Pipelines
31	Leather and Leather Products	47	Transportation Services
32	Stone, Clay, Glass, and Concrete Products	48	Communications
33	Primary Metal Industries	49	Utilities (Electric, Gas, Sewer)
34	Fabricated Metal Products	51	Nondurable Goods Wholesaling
35	Machinery	76	Miscellaneous Repair Services
36	Electrical and Electronic Machinery		

(NOTE: SIC numbers contain 4 digits, with the first two numbers signifying major groups as listed above.)

Therefore, to be subject to the provisions of ECRA, the place of business must be engaged in operations that involve hazardous substances or hazardous wastes *and* have an SIC number among those specified in the act.

NONAPPLICABILITY

Industrial establishments not subject to ECRA are those that are not involved in the generation, manufacturing, refining, transportation, treatment, storage, handling, or disposal of hazardous substances or wastes; or are not within Standard Industrial Classification major group numbers 22–39, 46–49, 51 and 76; or were closed, transferred, or sold prior to ECRA's effective date of December 31, 1983. However, if hazardous substances or wastes remain on the site in a storage vessel(s), surface impoundment(s), landfill(s), or another type of storage facility(s), even though the property was sold or closed before December 31, 1983, the operation is considered on-going by the department under ECRA and, therefore, all prior operations become subject to review under the ECRA legislation.

It is important to note that since retail gasoline stations have an SIC number of 55, **they are not subject to the provisions of ECRA**.

In addition, ECRA does not apply to those *portions* of industrial establishments currently subject to operational closure and postclosure maintenance requirements pursuant to the New Jersey Solid Waste Management Act, the New Jersey Major Hazardous Waste Facilities Siting Act, or the federal Solid Waste Disposal Act. It does not apply to any establishment engaged in the production and distribution of agricultural commodities. This only excludes individual or family farm operations engaged in the production and distribution of agricultural commodities and does not exclude establishments primarily engaged in the wholesale distribution of animal feeds, fertilization, agricultural chemicals, pesticides, seeds, and other farm supplies, except grains.

In summary, the place of business must have a primary SIC number as indicated above *and* must in some way handle hazardous substances or hazardous wastes. Any facility which fails either of these tests, or is specifically excluded, is not subject to ECRA.

LETTERS OF NONAPPLICABILITY

In some cases, various mortgage lenders, title insurers, or purchasers, will require written confirmation from the seller that a specific property or facility is not subject to ECRA. If you are in such a position, a letter confirming nonapplicability may be obtained from the Bureau of Industrial Site Evaluation by submitting a duly notarized affidavit clearly describing the reason that you believe makes the transaction not applicable.

The reasons for nonapplicability may include the absence of an industrial establishment with a SIC number addressed by the act, no involvement with hazardous substances or hazardous wastes on the property, the cessation of all operations (including the storage of hazardous substances such as fuel oil) prior to December 31, 1983, or other reasons supported by appropriate documentation. In the first case, the on-site activities must be described in sufficient detail for an independent interpretation of the appropriate SIC number to be made. In the second case, the affidavit must include a statement that the signer has reviewed the lists of hazardous substances and hazardous wastes and has determined that no such materials are generated, manufactured, refined, transported, treated, stored, or disposed on the property.

The statement must also provide evidence of the signer's qualifications to make such claims authoritatively—for example, a landlord generally would be unable to certify the absence of hazardous materials in a tenant's operation.

Any such affidavits should be sent to the Bureau of Industrial Site Evaluation, CN028, Trenton, New Jersey 08625, Attn: Applicability Section.

EXEMPTIONS

Subgroups or classes of operations within applicable SIC code numbers may be exempt provided the industrial establishments do not pose a risk to public health and safety. Such exemptions may be granted by the department. Industrial establishments must submit all appropriate documentation, evidence, and other proofs that they deem justify exemption as a subgroup or class from the act. Upon a finding that a subgroup, or classes of operations within relevant subgroups, does not pose a risk to the public health and safety, the department may amend N.J.A.C. 7:1–3.20(e) to exempt the subgroup or class of operations from consideration as an industrial establishment for the purpose of the act.

WHAT MUST INDUSTRIAL ESTABLISHMENTS DO? (See Section N.J.A.C. 7:1–3.7)

If you are unsure whether you are subject to ECRA, you may call the Bureau of Industrial Site Evaluation at (609) 633-7141 or 633-6690 for further clarification. If you have determined that you are subject to the act, you must submit an Initial Notice, forms for which are included within this package. Two separate application forms cover the Initial Notice requirements. The first application, referred to as the General Information Submission (GIS), is due within five days of the action which triggers ECRA. The GIS consists of readily available information such as name, address, SIC number of the industrial establishment, names, current addresses, and descriptions of all previous known operators and operations at the site; list of all federal and state permits for the site; list of any and all enforcement actions against the facility; and other necessary information noted on the application. The GIS must be submitted as written notice to the department within five days of the public release of the decision to close operations or within five days of the execution of an agreement of sale or any option to purchase.

The second part of the Initial Notice, known as the Site Evaluation Submission (SES), must be submitted no later than 30 days following your public release of the decision to close operations or execution of an agreement of sale or option to purchase.

The SES to be provided using the enclosed application is essentially an environment evaluation of the facility. It includes such items as a scaled site map identifying all areas where hazardous substances or wastes are located, a detailed description of the current operations, descriptions of the types and locations of hazardous waste storage facilities, a complete inventory of hazardous substances and wastes, a detailed sampling plan, a decontamination plan (if the facility is closing), and any other information deemed necessary by the department in order to evaluate the environmental conditions of the site.

The submittal of this information ultimately may result in the department's either approving a Negative Declaration or requiring the preparation and implementation of a Cleanup Plan. A Negative Declaration is a written affidavit duly notarized and signed by an authorized officer or management official of the industrial establishment stating that there has been no discharge of hazardous substances or wastes on the property or that any such discharge has been cleaned up in accordance with department-approved procedures. The affidavit should also state that no hazardous substances or wastes remain on the site above a level approved by the Department. No Negative Declaration or Cleanup Plan can be approved until the Initial Notice (GIS and SES) has been completed and all sample results have been submitted.

SAMPLING PLAN

The ECRA regulations require submission of detailed soil, ground water, and surface water sampling plans for department approval and implementation by the owner or operator of the industrial establishment prior to the submission of either a Cleanup Plan or a Negative Declaration. The sampling plan must include, but not be limited to, the following: a site map indicating sampling locations, sampling methodology, types of analyses, name of laboratory performing the analyses, and the Quality Assurance/Quality Control Plan. A more complete description of the required elements of the sampling plan is contained in the guidance package.

The owner or operator of an industrial establishment may propose that no sampling plan need be developed and implemented for the site; however, you must provide full justification, excluding economic reasons, for exemption from the sampling requirements.

INSPECTION

Department staff or its agents will inspect the industrial establishment based upon a review of the information submitted in the Initial Notice. The purpose of the inspection(s) will be to verify the environmental condition of the property described in the Initial Notice and, based on sampling results and visual observation, to decide whether a Negative Declaration or development and implementation of a Cleanup Plan is appropriate for the site.

It is worthwhile for the owner or operator to provide the department with access to all site areas, buildings, and records, since any delay during the review process eventually may postpone the closing, transfer, or sale of the property.

Further inspection may be necessary during both the sampling program and, if required, the cleanup program implementation to ascertain that these actions are carried out in conformance with the department-approved plans.

CLEANUP PLAN

A Cleanup Plan is a proposal developed by the owner or operator of the industrial establishment that describes in detail the method and operations that will be used to return the site to an environmentally acceptable condition. The Cleanup Plan must be approved by the department prior to implementation. It describes in detail the specific activities that the company or its contractor(s) will take to eliminate the presence of hazardous substances or waste which have been identified as needing remedial action. It also includes a description of the location, types, and quantities of hazardous substances and wastes that will remain on the premises, an evaluation of alternative cleanup methods, along with recommendations regarding the most practicable method of cleanup, detailed cost estimates, and a time schedule for the implementation of the cleanup plan. The sampling results from the detailed sampling program must be included in the cleanup plan.

The department will evaluate the plan, along with the other information submitted, to determine approval or denial of the proposal. If denied, the department will describe the deficiencies and the owner or operator will have to correct these deficiencies prior to department approval. If approved, the owner or operator has fourteen days in which to obtain a surety bond, letter of credit, or other financial security approved by the department guaranteeing the performance of the Cleanup Plan pursuant to the approval granted by the department. This assurance must be in the full amount estimated as necessary to complete the cleanup. The department then will conduct a final inspection of the site to ensure compliance with the Cleanup Plan.

For a detailed description of the financial requirements needed to insure proper cleanup plan implementation, consult N.J.A.C. 7:1–3.13 prior to undertaking your cleanup plan or cleanup activities. A model Wording of Instruments document for the financial assurance requirements has been developed by the department and can be obtained by contacting the Bureau of Industrial Site Evaluation.

Following approval of the company Cleanup Plan and receipt of appropriate financial assurances, the department may authorize the sale, transfer, or closure of the industrial establishment.

NONCOMPLIANCE WITH THE ACT

The act establishes strict penalties for noncompliance, the most compelling of which is the ability of the department or the transferee to void the sale or transfer of the industrial establishment. The transferee is also entitled to recover damages from the transferor plus render the owner or operator of the industrial establishment strictly liable, without regard to fault, for all cleanup and removal costs and for all direct or indirect damages resulting from the failure to implement the cleanup plan.

In addition, the law grants the state the authority to collect penalties of not more than $25,000 per day for falsifying information and/or general noncompliance with the act.

OFTEN-ASKED QUESTIONS

1. Where can I get a copy of the ECRA regulations?

The ECRA regulations are included with this package of material. Additional copies may be obtained by writing the Bureau of Industrial Site Evaluation, CN-028, Trenton, New Jersey 08625.

2. What forms do I use?

Included in this package are the application forms necessary to meet the ECRA Initial Notice requirements. The first part of the application is entitled the *General Information Submission* (GIS) and must be filled out and submitted to the Bureau of Industrial Site Evaluation, CN-028, Trenton, New Jersey 08625 within *five days* following the public release of the decision to close operations or execution of an agreement of sale or option to purchase. This part solicits only general information regarding site ownership, permits, etc. The second part of the application is more technical in nature and is entitled the *Site Evaluation Submission* (SES). This must be filled out and submitted to the bureau within *thirty days* following public release of the decision to close operations or execution of an agreement of sale or option to purchase. The SES must be submitted in duplicate.

3. What happens after I file an initial notice?

After both parts of the application forms are provided to the department and the Initial Notice is considered administratively complete, a preliminary inspection will be scheduled at which time you will be asked to walk through the property with the inspector. After this site inspection, the department will prepare a preliminary inspection report detailing site conditions and will provide you with guidance concerning ECRA compliance. The department will supplement all Initial Notice submissions with their own review of appropriate and available records of federal, state, county, and municipal agencies pertaining to the industrial establishment. The department may also conduct additional inspections and must be granted necessary access to all site areas, buildings, and records upon reasonable notice.

During the site inspection, the department's representative must be accompanied by appropriate technical, scientific and engineering representatives of the industrial establishment and must be granted access to all site areas, buildings, and records. After the site inspection, a decision on the acceptability of the sampling plan will be made.

Additional inspections may also be required.

4. Is a sampling plan needed in all cases?

This depends on the complexity of the site. A sampling plan is required by N.J.A.C. 7:1–3; however, the regulations also state that the owner or operator of an industrial establishment may propose that no sampling plan need be developed and implemented for the site. To do this, the applicant must provide full documentation of the justifications, other than economic reasons, for exemption from the requirements. This must be submitted to the department at the time of the Initial Notice for review and approval. The department will then make a determination of the need for sampling.

5. Can either sampling or a cleanup begin prior to department approval of the sampling or cleanup plan?

The department prefers that no sampling or cleanup begin until after a formal review of the proposal has been conducted and its completeness determined. We also recognize that many companies are hard pressed for time and are anxious to begin prior to DEP approval. Any sampling or cleanup work performed prior to an official approval by the department will be at the applicant's own risk, and the department may require additional sampling or cleanup depending on the methods and results of the unapproved programs.

6. When is sampling required?

This is a departmental judgment based on many factors including the history of the site, results of the ECRA and previous inspections, and any other information that may come forth showing that the property has been the site of spills or that the land was used to treat or dispose of wastes. The sampling program will be utilized to determine, along with other criteria such as the site inspection, whether a Negative Declaration will be approved or a full-scale cleanup is required prior to the sale, transfer, or closing of the property. In some cases, minor cleanups may be recommended by our representative during the inspection.

7. How long does it take the department to evaluate sampling results?

This depends on the type and number of analyses and the work load of the various bureaus involved in such reviews. Typically, it takes anywhere from three days or three weeks.

8. How can I get a list of laboratories capable of required analytical work?

This list can be obtained by contacting:

Mr. Jerry Bundy
Lab Certification Inspector
CN-402
Trenton, New Jersey 08625

or by calling (609) 292-3950.

9. What is a hazardous substance and where can I find the list?

The regulations define hazardous substances as those elements and compounds, including petroleum products, defined by the department after a public hearing, included on the "List of Hazardous Substances" found in Appendix

A of N.J.A.C. 7:1E. Appendix A includes a list of hazardous substances adopted by the Environmental Protection Agency pursuant to Section 311 of the Federal Water Pollution Control Act Amendments of 1972, 33 U.S.C. §1321, and a list of toxic pollutants designated by Congress or the Environmental Protection Agency pursuant to Section 307 of that Act, 33 U.S.C. §1317. Sewage and septage are not considered hazardous substances.

The regulations define hazardous waste as any amount of any waste substances required to be reported to the Department on the special waste manifest pursuant to N.J.A.C. 7:26–7.4, or designated as a hazardous waste pursuant to N.J.A.C. 7:26–8. The Department may expand, after public hearing, what constitutes a hazardous substance or waste through regulatory amendments.

The lists of hazardous substances and hazardous wastes are included in this package.

10. How does one get a deferral of implementation of an approved cleanup plan?

N.J.A.C. 7:1–3.14 outlines the procedures. If the premises of the industrial establishment would be subject to substantially the same use by the purchaser, transferee, mortgagee, or other party to the transfer, the owner or operator of the industrial establishment may apply, in writing, to the department for authorization to defer implementation of an approved Cleanup Plan until the use changes or until the purchaser, transferee, mortgagee, or other party to the transfer closes, terminates, or transfers operations.

The person applying for a deferral must prepare a written certification which is notarized and signed by an officer or management official stating the above and must be supported by similar documentation from the new owner or operator.

If this certification is denied by the department, immediate implementation of the approved Cleanup Plan shall be undertaken.

A Deferral can only be granted after a Cleanup Plan has been submitted and approved by the department. Requests prior to approval will be immediately denied as premature. Deferral of a Cleanup Plan will be granted only when it can be shown that the deferral will not cause or exacerbate an environmental pollution problem or represent a risk to the health of residents or employees.

11. Can the purchaser take responsibility for a cleanup?

Yes, provided the department is informed and *approves* the transfer of responsibility in writing.

12. How do I obtain an exemption from the ECRA requirements?

If you feel your industrial establishment should be exempt, you may petition the department in writing for an exemption as an individual or as a class from the requirements of the act. This exemption is based on a determination that the generic operation of the industrial establishment does not pose a risk to the public health and safety.

The owner or operator must submit appropriate documentation, evidence, and any other pertinent proof that you deem to justify an exemption from the ECRA review process.

The department on its own initiative may also establish a record based on experience or other appropriate research justifying an exemption. Upon a finding that a subgroup or class of operations within relevant subgroups do not pose a risk to the public health and safety, the department may amend N.J.A.C. 7:1–3.20(e) to exempt the subgroup or class of operations from consideration as an industrial establishment.

13. Can a cleanup determination be made based upon verbal communication of sampling results?

No, along with all Quality Assurance/Quality Control data, the written results must be submitted to this department prior to a determination of cleanup requirements.

14. Where can I find a list of environmental consultants?

The list of consultants who have responded to department requests for proposals on Superfund and N.J. Spill Fund work is included in this package.

Alternatively, many firms are listed in the Yellow Pages of telephone directories.

15. Can transactions be finalized before the ECRA review process is completed?

No, unless the applicant is willing to subject himself/herself to the risk of heavy penalties of as much as $25,000 per day and potential voiding of the transaction by either the department or the purchaser.

16. Is a warehouse that stores hazardous substances or wastes subject to ECRA?

If the warehouse is used to distribute nondurable goods such as but not limited to, paper, drugs, chemicals, petroleum, paints, and plastics, the warehouse falls within Major Group Number 51 of the Standard Industrial Classification numbering system and, therefore, is subject to ECRA.

17. How do I get the department to send me a letter certifying that my company/property is not subject to ECRA?

The department requires a notarized affidavit from the owner/operator of an industrial establishment specifying the company's SIC number and the activities that have been conducted at the premises which will dictate the appropriate SIC number and that you have familiarized yourself with those elements and compounds, including petroleum products, which are defined by the department as hazardous substances (Appendix A of N.J.A.C. 7:1E).

The affidavit should include the statement that the company/property is not engaged in operations which involve the generation, manufacture, refining, transportation, treatment, storage, or handling of hazardous substances or waste on the premises, above or below the ground, or that the company's SIC number is not among those subject to ECRA.

APPENDIX **H**

Cape Cod, Massachusetts, Model By-Laws and Health Regulations

- Underground Fuel and Chemical Storage Systems (4/1985)
- Toxic and Hazardous Materials (12/1981)
- Model Water Resource District Bylaw (6/1981)

CAPE COD PLANNING AND ECONOMIC DEVELOPMENT COMMISSION

MODEL HEALTH REGULATION TO PREVENT LEAKING OF UNDERGROUND FUEL AND CHEMICAL STORAGE SYSTEMS (Revised 4/85)

Under Chapter 111, Section 31 of the Massachusetts General Laws, the ________________________ Board of Health hereby adopts the following regulations to protect the ground and surface waters from contamination with liquid fuel or toxic materials from leaking storage tanks. The following regulations apply to all underground fuel and chemical storage systems.

Section 1. Definitions

ABNORMAL LOSS OR GAIN shall mean a loss or apparent gain in product exceeding 0.5 percent of the volume of product used or sold.

OPERATOR shall mean the lessee or person(s) in control of and having responsibility for the daily operation of the facility for the storage and dispensing of flammable and combustible liquids.

OWNER shall mean the person(s) who owns, as real property, the tank storage system used for the storage and dispensing of flammable and combustible liquids.

Section 2. Tank Registration

2.1 Every owner of an underground gasoline, fuel or chemical storage system shall file with the board of health the size, type, age and location of each tank, and the type of material stored, on or before ____________________. Evidence of date of purchase and installation, including fire department permit shall be included.

2.2 Owners of tanks for which evidence of installation date is not available shall at the order of the board of health have such tanks tested or uncovered for inspection. If in the opinion of the agent of the board of health or head of the fire department, the tank is not product tight, it shall be removed.

Section 3. Inventory Control

3.1 Every underground storage system shall have a method of accurately gauging the volume contained in the tank and a method of accurately metering the quantity of product removed during service. The metering device shall be maintained in accurate calibration. Storage systems in service at the time of passage of this regulation shall be in compliance within ninety days of the effective date of this regulation.

3.2 Accurate daily inventory records, as required by Massachusetts Fire Prevention Regulations-527 CMR 5.05 (3), shall be based on actual daily measurement and recording of tank product and water levels and the daily recording of actual sales, use and receipts. The inventory records shall include a daily computation of gain or loss. The mere recording of pump meter readings and product delivery receipts shall not constitute adequate inventory records.

3.3 The owner and operator shall participate in a program of regularly scheduled inventory verification. Frequency of inventory verification shall be as follows: annually, for systems from which less than 25,000 gallons/month of product is sold or used; semi-annually for systems for which 25,000–100,000 gallons/month is used or sold. Owners shall submit annually to the board of health a certified statement that inventory records have been maintained and reconciled as required by Massachusetts fire prevention regulations. Such records shall be made available to the board of health upon their request.

Where the storage tanks are owned by the operator, inventory verification shall be performed by a certified auditor or other independent qualified person approved by the board of health.

3.4 All steel tanks shall be subject to a Petro-Tite (Kent-Moore) Pressure Test or other pressure test providing equivalent safety and effectiveness 15 years after installation and annually thereafter. The 5 PSI air pressure test is no longer recognized by DEQE as a valid method of testing and will not be accepted by the ________________ Board of Health.

3.5 Nonconforming steel tanks installed prior to January 1, 1960 shall be removed and disposed of by ________________________. All other nonconforming steel tanks installed prior to the effective date of this regulation shall be removed when 20 years old. At such time the exhumed tank shall be examined for leaks. If a leak exists an investigation of amount and location of spilled substance shall be undertaken at the expense of the owner. If, in the opinion of the agent of the board of health, the spilled substance poses a significant threat to health and safety it shall be removed by the owner.

Section 4. Report of Leaks or Spills

4.1 Any person who is aware of a spill or abnormal loss of product shall report such spill or loss immediately to the head of the fire department, and within two hours to the board of health.

4.2 All leaking tanks *must be emptied within 24 hours of leak detection* and either removed or repaired within a time specified by the board of health, under the direction of the head of the fire department.

4.3 Service companies shall report to tank owners and the board of health any unaccounted for significant increase in heating fuel consumption which may indicate a leak.

4.4 All tank installations within four feet of high water table or within 100 feet of a surface water body shall be of fiberglass construction.

Section 5. Tank Selection and Installation

5.1 All tanks shall be properly installed as per Massachusetts Fire Prevention Regulations and manufacturers' specification, under the direction of the head of the fire department. Tanks shall be of approved design and protected from internal and external corrosion. The following tank construction systems are considered to provide adequate corrosion protection: all fiberglass construction; steel with bonded fiberglass or enamel coating and non-corrosive internal lining, and the Steel Tank Institute 3-Way Protection System. All underground storage of chemicals other than gasoline and fuel shall be contained in tanks approved by the agent of the board of health. Any other system must be shown to provide equivalent protection.

5.2 Tanks shall be installed by a manufacturer's approved installation contractor.

5.3 If it is necessary to replace or interior-coat an underground steel tank which developed a corrosion-induced leak, all other steel tanks at the facility which are the same age or older whether or not they are leaking shall be interior-coated or replaced with tanks that meet the requirements of 5.1.

5.4 If a cathodic protection system is installed, an ongoing monitoring and maintenance program shall be conducted. If sacrificial anodes have been installed, their proper operation shall be confirmed by a qualified person at least once a year.

5.5 The operator shall notify the head of the fire department prior to the commencement of tank installation. The head of the fire department, or the board of health may require repair of protective coatings prior to installation or final cover.

Section 6. Product Storage at Residential Sites

6.1 Following the effective date of this regulation the installation of underground fuel, gasoline or other chemical storage tanks on single-family or two-family residential sites is prohibited.

6.2 All fuel, gasoline or other chemical tanks in service at single-family or two-family residential sites on the effective date of this regulation shall be removed from service 30 years after the date of installation. If the date of installation is unknown, it shall be assumed to be January 1, 1960.

Section 7. Proximity to Water Supplies

7.1 The installation of subsurface fuel, gasoline or other chemical storage systems within 2,000 feet of a public water supply well is prohibited.

7.2 The board of health may require the installation of one or more groundwater observation wells at any site where fuel, gasoline or other chemical is stored underground within 2,000 feet of a public or private water supply well. Water samples from such observation wells may be required by the board of health at any reasonable time, and shall be analyzed at the expense of the owner at the order of the board of health.

Section 8. Costs

In every case the operator shall assume responsibility for costs incurred necessary to comply with this regulation.

Section 9. Variances

9.1 Variances from this regulation may be granted by the board of health after a hearing at which the applicant establishes that the installation or use of an underground storage tank will not adversely affect public or private water resources.

9.2 In granting variance, the board will take into consideration the direction of the ground water flow, soil conditions, depth to ground water, size, shape and slope of the lot, and existing and known future water supplies.

MODEL GENERAL BYLAW/HEALTH REGULATION TO CONTROL TOXIC AND HAZARDOUS MATERIALS IN THE TOWN OF ________________________ (December 1981)

Section 1. Findings

The Town of ________________________ finds that—

(1) The groundwater underlying this town is the sole source of its existing and future water supply, including drinking water;

(2) The groundwater aquifer is integrally connected with, and flows into, the surface waters, lakes, streams, and coastal estuaries which constitute significant recreational and economic resources of the town used for bathing and other water-related recreation, shellfishing, and fishing;

(3) Accidental spills and discharges of petroleum products and other toxic and hazardous materials have repeatedly threatened the quality of such groundwater supplies and related water resources on Cape Cod and in other Massachusetts towns, posing potential public health and safety hazards and threatening economic losses to the affected communities;

(4) Unless preventive measures are adopted to prohibit discharge of toxic and hazardous materials and to control their storage within the town, further spills and discharges of such materials will predictably occur, and with greater frequency and degree of hazard by reason of increasing construction, commercial and industrial development, population, and vehicular traffic in the Town of ________________________ and on Cape Cod;

(5) The foregoing conclusions are confirmed by findings set forth in the Environmental Impact Statement and *Water Quality Management Plan for Cape Cod* (September 1978), prepared by the Cape Cod Planning and Economic Development Commission pursuant to Section 208 of the Federal Clean Waters Act; by the report entitled *Chemical Contamination* (September 1979), Commonwealth of Massachusetts; and by the report, *Chemical Quality of Ground Water, Cape Cod, Massachusetts* (1979), prepared by the U.S. Geological Survey.

Section 2. Authority

The Town of ________________________ adopts the following measures under its home rule powers, its police powers to protect the public health and welfare, and its authorization under Chapter 40, M.G.L.A., Sec. 21. (FOR HEALTH REGULATION: under its authorization under Chapter 111, Sec. 31.)

Section 3. Definitions

(a) The term, "discharge" means the accidental or intentional spilling, leaking, pumping, pouring, emitting, emptying or dumping of toxic or hazardous material upon or into any land or waters of the Town of ____________________________. Discharge includes, without limitation, leakage of such materials from failed or discarded containers or storage systems, and disposal of such materials into any on-site sewage disposal system, drywell, catch basin, or unapproved landfill.

The term, "discharge" as used and applied in this bylaw, does not include the following:

(1) proper disposal of any material in a sanitary or industrial landfill that has received and maintained all necessary legal approvals for that purpose;
(2) application of fertilizers and pesticides in accordance with label recommendations and with regulations of the Massachusetts Pesticide Control Board;
(3) application of road salts in conformance with the Snow and Ice Control Program of the Massachusetts Department of Public Works; and
(4) disposal of "sanitary sewage" to subsurface sewage disposal systems as defined and permitted by Title 5 of the Massachusetts Environmental Code.

(b) The term, "toxic or hazardous material," means any substance or mixture of such physical, chemical, or infectious characteristics as to pose a significant actual or potential hazard to water supplies, or other hazard to human health, if such substance or mixture were discharged in this town. "Toxic or hazardous materials" include, without limitation, organic chemicals, petroleum products, heavy metals, radioactive or infectious wastes, acids, and alkalies, and include products such as pesticides, herbicides, solvents, and thinners. The following activities, without limitation, are presumed to involve the use of toxic or hazardous materials, unless and except to the extent that anyone engaging in such an activity can demonstrate the contrary to the satisfaction of the board of health:

- Airplane, boat, and motor vehicle service and repair
- Chemical and bacteriological laboratory operation
- Cabinet making
- Dry cleaning
- Electronic circuit assembly
- Metal plating, finishing, and polishing
- Motor and machinery service and assembly
- Painting, wood preserving, and furniture stripping
- Pesticide and herbicide application
- Photographic processing
- Printing

The Board of Health may, consistent with this definition, and by authority of Chapter 111, Section 31, issue regulations further identifying specific materials and activities involving the use of materials which are toxic or hazardous.

Section 4. Prohibitions

(a) The discharge of toxic or hazardous materials within the Town of ______________________ is prohibited.

(b) Outdoor storage of toxic or hazardous materials is prohibited, except in product-tight containers which are protected from the elements, leakage, accidental damage, and vandalism, and which are stored in accordance with all applicable requirements of Section 5 of this bylaw. For purposes of this subsection, road salts and fertilizers shall be considered as hazardous materials.

Section 5. Storage Controls, Registration, and Inventory

(a) Except as exempted below, every owner, and every operator other than an owner, of a site at which toxic or hazardous materials are stored in quantities totalling at any time more than ____________ gallons liquid volume or ____________ pounds dry weight, shall register with the board of health the types and quantities of materials stored, location, and method of storage. The board of health may require that an inventory of such materials be maintained on the premises and be reconciled with purchase, use, sales, and disposal records on a monthly basis, in order to detect any product loss. Registration required by this subsection shall be submitted within 60 days of the effective date of this bylaw, and annually thereafter. Maintenance and reconciliation of inventories shall begin within the same 60-day period. Exemptions: Registration and inventory requirements shall not apply to the following:

   (1) Fuel oil stored in conformance with Massachusetts Fire Prevention Regulations and regulations of the ______________________ Board of Health for the purpose of heating buildings located on the site; or

   (2) The storage of toxic and hazardous materials at a single-family or two-family dwelling, except where such materials are stored for use associated with a professional or home occupation use as defined by Section ______________ of the Zoning Bylaws of the Town of ______________________.

(b) Toxic or hazardous wastes shall be held on the premises in product-tight containers and shall be removed and disposed of in accordance with the Massachusetts Hazardous Waste Management Act, Ch. 21C, MGLA.

(c) The board of health may require that containers of toxic or hazardous materials be stored on an impervious, chemical-resistant surface compatible with the material being stored, and that provisions be made to contain the product.

Section 6. Report of Spills and Leaks

Every person having knowledge of a spill, leak, or other loss of toxic or hazardous materials believed to be in excess of ____________ gallons shall immediately report the spill or loss of same to the board of health or other public safety official.

Section 7. Enforcement

(a) The provisions of this bylaw shall be enforced by the board of health. The agent of the board of health may, according to law, enter upon any premises at any reasonable time to inspect for compliance.
(b) Upon request of an agent of the board of health, the owner or operator of any premises at which toxic or hazardous materials are used or stored shall furnish all information required to enforce and monitor compliance with this bylaw, including a complete list of all chemicals, pesticides, fuels, and other toxic or hazardous materials used or stored on the premises, a description of measures taken to protect storage containers from vandalism, corrosion, and spillage, and the means of disposal of all toxic or hazardous wastes produced on the site. A sample of wastewater disposed to on-site septic systems, drywells, or sewage treatment systems may be required by the agent of the board of health.
(c) All records pertaining to storage, removal, and disposal of toxic or hazardous materials shall be retained by the owner or operator for no less than three years, and shall be made available for review upon request of the agent of the board of health.
(d) Certification of conformance with the requirements of this bylaw by the board of health shall be required prior to issuance of construction and occupancy permits for any nonresidential uses.

Section 8. Violation

Written notice of any violation of this bylaw shall be given to the owner and operator by the agent of the board of health, specifying the nature of the violation; and corrective measures that must be undertaken, including containment and cleanup of discharged materials; and preventive measures required for avoiding future violations; and a schedule of compliance. Requirements specified in such a notice shall be reasonable in relation to the public health hazard involved and the difficulty of compliance. The cost of containment and cleanup shall be borne by the owner and operator of the premises.

Section 9. Penalty

Penalty for failure to comply with any provisions of this bylaw shall be $200.00 per day of violation, after notice thereof under Section 8, above. (FOR HEALTH REGULATION: shall be $20.00)

Section 10. Severability

Each provision of this bylaw shall be construed as separate, to the end that if any part of it shall be held invalid for any reason, the remainder shall continue in full force and effect.

MODEL WATER RESOURCE DISTRICT BYLAW (June 22, 1981) CAPE COD PLANNING AND ECONOMIC DEVELOPMENT COMMISSION

A. Findings

The Town of ______________________ finds that—

1. The groundwater underlying this town is the sole source of its existing and future water supply, including drinking water;
2. The groundwater aquifer is integrally connected with, and flows into, the surface waters, lakes, streams, and coastal estuaries which constitute significant recreational and economic resources of the town used for bathing and other water-related recreation, shellfishing, and fishing;
3. Accidental spills and discharges of petroleum products and other toxic and hazardous materials have repeatedly threatened the quality of such groundwater supplies and related water resources on Cape Cod and in other Massachusetts towns, posing potential public health and safety hazards and threatening economic losses to the affected communities;
4. Unless preventive measures are adopted to prohibit discharge of toxic and hazardous materials and to control their storage within the town, further spills and discharges of such materials will predictably occur, and with greater frequency and degree of hazard by reason of increasing construction, commercial and industrial development, population, and vehicular traffic in the Town of ______________________ and on Cape Cod;
5. The foregoing conclusions are confirmed by findings set forth in the Environmental Impact Statement and *Water Quality Management Plan for Cape Cod* (September 1978), prepared by the Cape Cod Planning and Economic Development Commission pursuant to Section 208 of the Federal Clean Waters Act; by the report entitled *Chemical Contamination* (September 1979), prepared by the Special Legislative Commission on Water

(Note: FOR HEALTH REGULATION: The words "this bylaw" should read "this regulation" in all cases.)

Supply, Commonwealth of Massachusetts; and by the report, *Chemical Quality of Ground Water, Cape Cod, Massachusetts* (1979), prepared by the U.S. Geological Survey.

B. Water Resource District

The objective of the water resource district is to protect the public health by preventing contamination of the ground and surface water resources providing water supply for the Town.

C. Use Regulations

1. Prohibited Uses. Within the water resources district the following uses are prohibited: sanitary landfills, junk yards, municipal sewage treatment facilities with on-site disposal of primary or secondary-treated effluent, car washes, road salt stockpiles, dry cleaning establishments, boat and motor vehicle service and repair, metal plating, chemical and bacteriological laboratories, and any other use which involves as a principal activity the manufacture, storage, use, transportation, or disposal of toxic or hazardous materials, except as allowed by special permit below.
2. Special Permit Uses. Within the Water Resources District the following shall be allowed only if granted a Special Permit:
   a. Transportation terminals.
   b. Any principal use involving the sale, storage, transportation of fuel oil or gasoline.
   c. Any use involving the retention of less than 30% of lot area in its natural state with no more than minor removal of existing trees and ground vegetation, or rendering impervious more than 40% of lot area.
   d. Any use involving on-site disposal of process wastes from operations other than personal hygiene and food for residents, patrons, and employees.
   e. Any use (other than a single-family dwelling) with a sewage flow, as determined by Title 5 of the State Environmental Code, exceeding 110 gallons per day per 10,000 sq. ft. of lot area or exceeding 15,000 gallons per day regardless of lot area.
   f. Any use involving the generation of toxic or hazardous materials in quantities greater than associated with normal household use.

D. Special Permits

1. Special Permit Granting Authority. The special permit granting authority (SPGA) shall be the ________________ except that where a special permit is required by Section ____________, the SPGA authorized by that section shall be the SPGA for the special permit under this section. Such special permit shall be granted if the SPGA determines, in conjunction

with other town agencies as specified in Section D.2 below, that the intent of this bylaw as well as its specific criteria are met. In making such determination, the SPGA shall give consideration to the simplicity, reliability, and feasibility of the control measures proposed and the degree of threat to water quality which would result if the control measures failed. The SPGA shall explain any departures from the recommendations of the other town agencies in its decision.

2. Review by Other Town Agencies. Upon receipt of the special permit application, the SPGA shall transmit one copy each to ______________________ for their written recommendations. Failure to respond in writing within 30 days shall indicate approval by said agencies. The necessary number of copies of the application shall be furnished by the applicant.

3. Special Permit Criteria. Special permits under Section __________ shall be granted only if the SPGA determines, in conjunction with other town agencies as specified above, that groundwater quality resulting from on-site waste disposal and other on-site operations will not fall below federal or state standards for drinking water, or, if existing groundwater quality is already below those standards, on-site disposal will result in no further deterioration.

4. Submittals. In applying for a special permit under this section, the information listed below shall be submitted as specified in Section __________.

   a. A complete list of all chemicals, pesticides, fuels, and other potentially toxic or hazardous materials to be used or stored on the premises in quantities greater than those associated with normal household use, accompanied by a description of measures proposed to protect all storage containers/facilities from vandalism, corrosion, and leakage, and to provide for control of spills.

   b. A description of potentially toxic or hazardous wastes to be generated, indicating storage and disposal methods.

   c. Evidence of approval by the Massachusetts Department of Environmental Quality Engineering (DEQE) of any industrial waste treatment or disposal system or any wastewater treatment system over 15,000 gallons per day capacity.

   d. For underground storage of toxic or hazardous materials, evidence of qualified professional supervision of system design and installation.

   e. Analysis certifying compliance with Subsection D.3. of Section ________________ such analysis to be done by a technically qualified expert.

E. Design and Operations Guidelines

Except for single-family dwellings, the following design and operation guidelines shall be observed within the Water Resource District.

1. Safeguards. Provision shall be made to protect against toxic or hazardous materials discharge or loss resulting from corrosion, accidental damage, spillage, or vandalism through measures such as: spill control provisions in the vicinity of chemical or fuel delivery points; secured storage areas for toxic or hazardous materials; and indoor storage provisions for corrodable or dissolvable materials. For operations which allow the evaporation of toxic or hazardous materials into the interiors of any structures, a closed vapor recovery system shall be provided for each such structure to prevent discharge of contaminated condensate into the groundwater.
2. Location. Where the premises are partially outside of the Water Resource District, potential pollution sources such as on-site waste disposal systems shall be located outside the District to the extent feasible.
3. Disposal. For any toxic or hazardous wastes to be produced in quantities greater than those associated with normal household use, the applicant must demonstrate the availability and feasibility of disposal methods which are in conformance with Ch. 21C, MGLA.
4. Drainage. All runoff from impervious surfaces shall be recharged on the site, diverted towards areas covered with vegetation for surface infiltration to the extent possible. Dry wells shall be used only where other methods are not feasible, and shall be preceded by oil, grease, and sediment traps to facilitate removal of contaminants.

F. Violations

Written notice of any violation of Section ____________ shall be provided by the Building Inspector to the owner of the premises, specifying the nature of the violations and a schedule of compliance, including cleanup of any spilled materials. This compliance schedule must be reasonable in relation to the public health hazard involved and the difficulty of compliance. In no event shall more than 30 days be allowed for either compliance or finalization of a plan for longer-term compliance.

G. Definition

TOXIC OR HAZARDOUS MATERIALS. Any substance or mixture of such physical, chemical, or infectious characteristics as to pose a significant actual or potential hazard to water supplies, or other hazard to human health, if such substance or mixture were discharged to land or waters of this town. Toxic or hazardous materials include radioactive or infectious wastes, acids, and alkalies, and include products such as pesticides, herbicides, solvents, and thinners. Wastes generated by the following activities, without limitation, are presumed to be toxic or hazardous, unless and except to the extent that anyone engaging in such an activity can demonstrate the contrary to the satisfaction of the board of health:

- Airplane, boat, and motor vehicle service and repair

- Chemical and bacteriological laboratory operation
- Cabinet making
- Dry cleaning
- Electronic circuit assembly
- Metal plating, finishing, and polishing
- Motor and machinery service and assembly
- Painting, wood preserving, and furniture stripping
- Photographic processing
- Printing

APPENDIX I
Biographical Sketches of Committee Members

JEROME B. GILBERT (Chairman) holds a B.S. in civil engineering and an M.S. in civil engineering from Stanford University and is General Manager of the East Bay Municipal Utility District, Oakland, California. His experience in environmental engineering includes the planning and financial, institutional, and regulatory aspects of water management. He has been Executive Officer for the California State Water Resources Control Board, and a principal of Brown & Caldwell and J. B. Gilbert & Associates Consulting Engineers. He has been Chairman of the San Francisco Bay Regional Water Quality Control Board and Chairman of the California Health Effects Advisory Committee on reuse of reclaimed waste water. He has held American Water Works Association offices including President and international technical representative. He has served on the Water Science and Technology Board of the National Research Council and its Potomac River Committee.

EULA BINGHAM received her Ph.D. in zoology from the University of Cincinnati in 1958. She was formerly Assistant Secretary of Labor for the Occupational Safety and Health Administration from 1977 to 1981. Currently, Dr. Bingham is Vice President and University Dean for Graduate Studies and Research at the University of Cincinnati, where she has been since 1961.

JOHN J. BOLAND holds a B.A. in electrical engineering, an M.A. in governmental administration, and a Ph.D. in environmental economics. He is a registered professional engineer. His background includes management positions in water/waste water utilities, consulting activities at all levels of government and in private industry, teaching, and research. He is currently Professor of Geography and Environmental Engineering at Johns Hopkins University. Dr. Boland has published widely on economic aspects of water and resource policy. He is an associate editor of *Water Resources Research* and chairman of the Economic Research Committee of the American Water Works Association. He has served on a number of committees and panels of the National Research Council and is currently chairman of the Water Science and Technology Board.

ANTHONY D. CORTESE received his B.S. and M.S. degrees in environmental engineering from Tufts University and a Sc.D. in environmental health sciences from Harvard University. He is currently Director of Tufts University's Center for Environmental Management, an interdisciplinary research, education, and policy center dealing with hazardous wastes and toxic substances. From 1979 to 1984 he was Commissioner of the Massachusetts Department of Environmental Quality Engineering. In that position he managed all hazardous and solid waste management, water supply, water pollution, wetlands and ground water protection, and air pollution control activities for the state of Massachusetts. Prior to 1979, he held a variety of scientific and management positions with the state of Massachusetts and the U.S. Environmental Protection Agency—all related to water, air, and land quality management.

THOMAS M. HELLMAN obtained a B.A. in chemistry from Williams College and a Ph.D. in organic chemistry from Pennsylvania State University. Dr. Hellman has 16 years of experience as an environmental scientist and manager in industry. From 1970 to 1973, he worked for Union Carbide and simultaneously taught at West Virginia University. In 1973, he joined Allied Corporation, holding various positions in several geographical locations including Manager for Air and Water Programs and Department Head for Health, Safety and Environmental Sciences. In July 1984, he moved to the position as Director of Allied's legislative and regulatory affairs pertaining to environmental matters in Washington, D.C. In 1985, Dr. Hellman joined General Electric Company as Corporate Manager of Health, Safety and Environmental Protection. Dr. Hellman is the past Chairman of the Chemical Manufacturers Association's Environmental Management Committee (1984–1985). Also, he served for four years on the New Jersey Hazardous Waters Advisory Council.

WILEY HORNE received from the University of North Carolina at Chapel Hill a B.S. in mathematics in 1967, an M.S. in chemistry and biology in 1969, and a Ph.D. in water resources engineering in 1972. From 1972 to 1976, he was planning engineer, and subsequently head of facilities planning, with the Los Angeles County Sanitation District. From 1977 to 1982, he was project manager of the Orange and Los Angeles counties water reuse study. At present, he is Chief, Local Projects Branch, Metropolitan Water District of Southern California, where he is in charge of development of waste water reclamation, surface/ground water conjunctive use, and desalting projects. Recently, Dr. Horne managed a regional study for ground water protection planning for southern California.

HELEN INGRAM received her undergraduate education at Oberlin College and earned her Ph.D. in political science at Columbia University in 1967. She taught at the University of New Mexico for several years, worked on the staff of the National Water Commission, was a research fellow at Resources for the Future, and has been on the staff of the University of Arizona since 1972, where she is now Professor of Political Science. Dr. Ingram has done considerable research on public policy aspects of water resources, including ground water management, and has authored books and journal articles in that field. She is also a member of the Water Science and Technology Board.

THOMAS M. JOHNSON obtained his B.A. from Augustana College in geology and holds M.S. degrees in geology and in water resources management, both from the University of Wisconsin. He is currently completing requirements for his Ph.D. at the University of Illinois. From 1975 to 1986, Mr. Johnson worked for the Illinois State Geological Survey, where he was Head of the Ground Water Section. Recently, Mr. Johnson became Principal Hydrogeologist and Manager of the Hydrogeology Group of Levine-Fricke Consulting Engineers and Hydrogeologists, Oakland, California. He is regarded as an expert in various aspects of understanding ground water contamination in the vadose (or unsaturated) zone—including monitoring and modeling. In addition, Mr. Johnson is a member of the faculty for the National Water Well Association's series of short courses on ground water monitoring and corrective actions for contaminated ground water.

SUE LOFGREN received her B.A. at the University of Phoenix in management. As a partner in the firm, the Forum, which designs and implements programs geared to increasing public awareness of natural resources and other public policy issues, she specializes in conflict management particularly relating to water issues. She has conducted seminars in water quality management, water reuse, water law, and ground water protection management. She has served on many advisory committees such as the EPA's National Drinking Water Advisory Council, where she chaired the Public Awareness Sub-Group and was a member of the Reuse and Underground Injection Sub-Groups. At present, she is Chairman of the regional Ground Water Users Advisory Council, Department of Water Resources, and also the Regional Water Quality Policy Advisory Committee. She participated in the drafting of the Arizona Ground Water Protection Regulations.

PAULA MAGNUSON received her B.A. from Smith College in environmental biology. From 1981 to present she has been a Senior Environmental Scientist at Geraghty & Miller, ground water management consultants in New York. Ms. Magnuson is responsible for monitoring the development of federal and state ground water protection programs for Geraghty & Miller. She has served as a consultant to industrial clients concerning petroleum and chemical cleanup programs. She has served on numerous state advisory committees and has participated extensively in public participation programs for ground water protection controls. Ms. Magnuson is also Chairman of the Ground Water Protection Committee of the National Water Well Association. Prior to 1981, she worked for about 10 years for a county government on Cape Cod in planning and implementing ground water protection plans, among other things.

PERRY L. McCARTY obtained his Sc.D. in sanitary engineering from MIT in 1959. He has been a faculty member of the Department of Civil Engineering at Stanford University since 1962, and from 1979 to 1985 was Departmental Chairman. From 1962 to 1967 he was Associate Professor, from 1967 to 1975 he was Professor, and since that time has served as Silas H. Palmer Professor of Civil Engineering. He has been involved with ground water recharge and water quality studies for several years in California and elsewhere. Dr. McCarty is a member of the National Academy of Engineering and has either chaired or been a member of many NRC boards and committees.

DWIGHT F. METZLER received his B.S. in civil engineering from the University of Kansas and his M.S. from Harvard University in sanitary engineering and worked for approximately 40 years for the states of Kansas and New York in the technical and management aspects of water resource, public health, and environmental programs. For 8 years Mr. Metzler directed New York's environmental program, first as Deputy Commissioner of the New York State Department of Health and then as Deputy Commissioner of the New York Department of Environmental Conservation. As Secretary he organized the new Department of Health and Environment for Kansas in 1974 and served under two governors. He retired from the position of Chief of water systems development in Kansas. Mr. Metzler has received many honors for his achievements in environmental engineering and public health. He is a member of the National Academy of Engineering.

CHRISTINE SHOEMAKER obtained her Ph.D. in mathematics from the University of Southern California in 1971, when she joined the Cornell faculty. Currently, she is Professor and Chair of the Department of Environmental Engineering at Cornell. In 1976 to 1977 she was a Visiting Assistant Systems Ecologist in the Department of Entomological Sciences at the University of California, Berkeley. Dr. Shoemaker has been a panel member for an NRC committee on pest control and is currently a member of the Environmental Studies Board's Scientific Council on Problems of the Environment. She is also a member of the FAO Expert Panel on Pest Management. Dr. Shoemaker was Co-Project Director of the alfalfa project in the Consortium for Integrated Pest Management (CIPM), an interdisciplinary, nationwide program sponsored by USDA and EPA. Her research involves the application of optimization, statistical, and mathematical analysis to environmental problems. She has published research articles on applications of mathematical modeling to integrated pest management, water supply and wastewater networks, and effects of acid precipitation.

DAVID A. STEPHENSON received his Ph.D. in hydrogeology in 1965 from the University of Illinois. His B.A. and M.S. degrees are in geology from Augustana College and Washington State University, respectively. He was a Professor of Hydrogeology and Director of the graduate degree program in Water Resources Management at the University of Wisconsin-Madison from 1965 to 1979. He also was Chief of the Water Resources Section of the Wisconsin Geological and Natural History Survey. He conducted research on ground water quality and on ground water flow systems, with particular emphasis on ground water/lake relationships. He also led initial efforts to develop a state ground water protection program by organizing legislative, Wisconsin Department of Natural Resources, university, and agency communication and planning activities. Dr. Stephenson was formerly Director of The Water Resources Group of Woodward-Clyde Consultants Western Region office and is now an Associate with Dames and Moore Consultants in Phoenix.

JAMES T.B. TRIPP obtained his L.L.B. from the Yale Law School in 1966 along with an M.S. in philosophy from Yale's graduate school. He was Assistant U.S. Attorney for Southern New York State from 1968 to 1973. Since 1973 he has been with the Environmental Defense Fund as head of its eastern water resources and land use program. His interests include wetlands protection and ground water quality protection.

Ex-Officio

DAVID W. MILLER received a B.A. and an M.S. in geology from Colby College and Columbia University in 1953. He is a well-known expert in applied ground water hydrology, and his career includes service with the U.S. Geological Survey and with a consulting firm that bears his name. He is skilled in both ground water quantity and quality disciplines. Mr. Miller is currently a member of the Water Science and Technology Board and was also member of the NRC Committee to Review the Metropolitan Washington Area Water Supply Study.

Technical Consultant

JOHN B. ROBERTSON obtained his M.S. degree in geological engineering from Colorado School of Mines. From 1961 to 1984 he worked as a research and supervisory ground water hydrologist with the Water Resources Division of the U.S. Geological Survey. Currently, he is employed with Roy F. Weston Consultants. His field of expertise is in ground water hydrology and behavior and the fate of contaminants in ground water.

Index

F

G

H

I

K

L

Q

R

S